Translating and Communicating Environmental Cultures

Environmental translation studies has gained momentum in recent years as a new area of research underscored by the need to communicate environmental concerns and studies across cultures. The dissemination of translated materials on environmental protection and sustainable development has played an instrumental role in transforming local culture and societies. This edited book represents an important effort to advance environmental studies by introducing the latest research on environmental translation and cross-cultural communication. Part I of the book presents the newest research on multilingual environmental resource development based at leading research institutes in Europe, Latin America, North America, and the Asia-Pacific. Part II offers original, thought-provoking linguistic, textual, and cultural analyses of environmental issues in genres as diverse as literature, nature-based tourism promotion, environmental marketing, environmental documentary, and children's reading. Chapters in this book represent original research authored by established and mid-career academics in translation studies, computer science, linguistics, and environmental studies around the world. The collection provides engaging reading and references on environmental translation and communication to a wide audience across academia.

Meng Ji is Associate Professor of Translation Studies at the School of Languages and Cultures, The University of Sydney, Australia.

Routledge Studies in Empirical Translation and Multilingual Communication

Series Editor: Meng Ji, *The University of Sydney, Australia*

Empirical Translation Studies (ETS) represents a rapidly growing field of research which came to the fore in the 1990s. From the early tentative use of computerised translation research tools to the systematic investigation of large-scale translation corpora using advanced statistical methods, ETS has made substantial progress in the development of solid empirical research methodologies which lie at the heart of the development of the field. There is a growing volume of research pursued in ETS as corpus translation studies has become a core component of translation studies at the postgraduate and research levels. To offer an appropriate and much-needed outlet for high-quality research in ETS, this book series selects and publishes latest translation research from around the world, in which the innovative use of corpus materials and related methodologies is essential. An important shared feature of these titles is their original contribution made to the advancement of empirical methodologies in translation studies which includes, but is not limited to, the corpus analysis, modelling and interpretation of translation corpora.

Health Translation and Media Communication
A Corpus Study of the Media Communication of Translated Health Knowledge
Meng Ji

Multicultural Health Translation, Interpreting and Communication
Edited by Meng Ji, Mustapha Taibi, and Ineke H.M. Crezee

Cross-Cultural Health Translation
Exploring Methodological and Digital Tools
Edited by Meng Ji

Translating and Communicating Environmental Cultures
Edited by Meng Ji

For more information about this series, please visit www.routledge.com/ Routledge-Studies-in-Empirical-Translation-and-Multilingual-Communication/ book-series/RSET

Translating and Communicating Environmental Cultures

Edited by Meng Ji

Routledge
Taylor & Francis Group

LONDON AND NEW YORK

First published 2020
by Routledge
2 Park Square, Milton Park, Abingdon, Oxon OX14 4RN

and by Routledge
52 Vanderbilt Avenue, New York, NY 10017

Routledge is an imprint of the Taylor & Francis Group, an informa business

First issued in paperback 2021

British Library Cataloguing-in-Publication Data
A catalogue record for this book is available from the British Library

Library of Congress Cataloging-in-Publication Data
Names: Ji, Meng, editor.
Title: Translating and communicating environmental cultures / edited by Meng Ji.
Description: New York, NY: Routledge, [2019] | Series: Routledge studies in empirical translation and multilingual communication |
Includes bibliographical references and index.
Identifiers: LCCN 2019003070 | ISBN 9781138359819 (hardback) |
ISBN 9780429433498 (ebook)
Subjects: LCSH: Ecology–Translating. | Environmental protection–Translating.
Classification: LCC QH541.187 .T73 2019 | DDC 577–dc23
LC record available at https://lccn.loc.gov/2019003070

ISBN: 978-1-138-35981-9 (hbk)
ISBN: 978-1-03-209108-2 (pbk)
ISBN: 978-0-429-43349-8 (ebk)

Typeset in ITC Galliard
by Newgen Publishing UK

Contents

Figures

Tables

Contributors

Nic Bertrand is with the Centre for Ecology & Hydrology, Lancaster Environmental Centre, UK.

David Blankman is the Chair of the ILTER's Information Management Committee (International Long Term Ecological Research), Israel.

Silvia Bruti is Associate Professor at the University of Pisa, Italy.

Maria Cristina Caimotto is Assistant Professor at the University of Torino, Italy.

Sabina Di Franco is Researcher at the National Research Council of Italy.

Pamela Faber Benitez is Chair Professor of Translation and Interpreting at the University of Granada, Spain.

Diego Ferreyra is Research fellow, Universidade Federal de Minas Gerais, Brazil. CONICET (Consejo Nacional de Investigaciones Científicas y Técnicas) – Argentina.

Xuebing Guo is with the Chinese Academy of Sciences, Beijing, China.

Honglin He is with the Chinese Academy of Sciences, Beijing, China.

Don Henshaw is with the Corvallis Forestry Sciences Laboratory, USA.

Stefan Jensen is Head of the European Environment Agency, Denmark.

Karpjoo Jeong is Professor at the Department of Internet and Multimedia Engineering, Konkuk University, Seoul, South Korea.

Meng Ji is Professor of Translation Studies, University of Sydney, Australia.

Yunlong Jia is Senior Corpus Research Scientist and CEO at HugeMind (Beijing) Education & Technology Co., Beijing, China.

Eun-Shik Kim is Professor at Department of Forestry Environment and Systems, College of Forest Science, Kookmin University, Seoul, South Korea.

Pilar León Araúz is a postdoctoral research fellow at the University of Granada, Spain.

Chau-Chin Lin is a research scientist at Taiwan Forestry Research Institute (TFRI), Taipei, Taiwan.

Sheng-Shan Lu is with the Taiwan Forestry Research Institute, Taiwan.

Sofia Malamatidou is Lecturer at the University of Birmingham, UK.

Elena Manca is Researcher and Lecturer at the University of Salento, Italy.

Margaret O'Brien is a Specialist at the Marine Science Institute, University of California, USA.

Takeshi Osawa is with the National Institute for Agro-Environmental Sciences, Ibaraki, Japan.

Éamonn Ó Tuama is a Senior Programme Officer at the GBIF Secretariat, Copenhagen, Denmark.

Adriana S. Pagano is Professor of Translation Studies at Universidade Federal de Minas Gerais, Brazil.

Paolo Plini is Researcher at the National Research Council of Italy.

John H. Porter is Research Associate Professor at the University of Virginia, USA.

Arianne Reimerink is Research fellow, Department of Translation and Interpreting, University of Granada.

André L. Rosa Teixeira is Assistant Professor at Universidade Federal de Minas Gerais, Brazil.

Davi Seabra Grossi is Research fellow, Universidade Federal de Minas Gerais, Brazil.

Mark Seligman is President of Spoken Translation Inc., Berkeley, USA.

Wen Su is with the Chinese Academy of Sciences, Beijing, China.

Kristin Vanderbilt is Associate Research Professor at University of New Mexico, USA.

Jiajin Xu is Professor of Corpus Linguistics at the Beijing Foreign Studies University, China.

Haibo Yang is Professor at the School of Chemistry and Molecular Engineering, East China Normal University, Shanghai, China.

Preface

How new concepts and social values are constructed and diffused within and across languages and cultures represents an intriguing research topic in the rapidly growing field of cross-cultural environmental studies. The promotion and dissemination of global environmental policies has given rise to an array of localised and culturally adapted environmental cultures and discourses.

New environmental idea sets and related social initiatives are not only revisiting and challenging environmental values within the cultural and knowledge systems of the target societies but also are increasingly interacting and networking with other local environmental cultures and social innovation efforts. Studies of environmental cultures make significant contributions to national and regional sustainable growth and development.

Translations of environment play an instrumental role in shaping the understanding of the relationship between human and environment across languages and cultures in historical and contemporary times. This book shows that translation has played and continues to play an instrumental role in the development of localised environmental and sustainable cultures and discourses around the world.

This book inquires and provides systematic profiling of key features, dimensions and processes of the translation, diffusion, adaptation, and development of environmental cultures and discourses. It focuses on the temporality, spatiality, cultural specificity, and cross-cultural comparability of local environmental cultures and practices.

The highly dynamic nature of environmental cultures can be explored productively through empirical linguistic investigation, for example, data-driven statistical modelling of multilingual environmental data collected from original and translated resources. This book explores interdisciplinary methodologies for environmental translation and communication studies by integrating research methodologies from translation studies, corpus linguistics, and quantitative or statistical modelling of multilingual digital environmental media data.

Chapters in this book represent original studies by using either qualitative environmental cultural studies or theoretically informed statistical modelling and analysis of multilingual environmental media data to reveal underlying patterns in the diffusion and development of new environmental cultural concepts, values, and social practices across language and cultures.

This book offers a global perspective of the growing environmental cultures in East Asia, Latin America, and Mediterranean Europe. The study of the development of environmental cultures provides an opportunity to analyse and compare national and sectoral efforts at driving sustainable development and growth in distinct social and cultural systems.

The rich insights to be gained in the study of environmental cultures and their interconnectivity around the world will help advance the new interdisciplinary research field of environmental translation of important social and environmental significance.

Part I
Multilingual environmental resources development

1 Translating environmental texts with EcoLexiCAT

Pilar León Araúz, Arianne Reimerink, and Pamela Faber Benitez

1.1 Introduction

In today's world, machine translation (MT) and computer-assisted translation (CAT) are a consolidated part of the professional translation workflow. Nevertheless, terminology management is still considered complex and time-consuming and often is not seamlessly integrated into the translation process. Furthermore, most terminological tools do not take into account the real search behaviour of end users such as translators (Tudhope et al., 2006; Durán Muñoz, 2012: 78) and most terminological modules in CAT tools do not go beyond a simple glossary of source and target terms. Apart from that, access to corpora is generally not provided in most CAT tools, despite the valuable phraseological information that a corpus can provide. An exception to this is the recently added Sketch Engine plug-in (available from the SDL AppStore) in SDL Trados Studio but, generally speaking, loss of translation quality and precious time are the inevitable consequences.

An excellent example of how to improve on the current situation is the initiative carried out by the TaaS project.[1] TaaS (Terminology as a Service) is a European project developed by a group of institutions and companies in the translation technology field who conceive 21st-century terminology in a user-friendly, collaborative, cloud-based environment (Gornostay, 2014). Their aim is to create a platform for instant access to the most up-to-date terms and for user participation in the acquisition, sharing, and reuse of multilingual terminological data. TaaS targets all types of language professionals but specifically focuses on translators as end users, as it provides the following terminology services: (1) automatic extraction of term candidates; (2) automatic recognition of translation equivalents in different public and industry terminology databases; (3) automatic acquisition of translation equivalents for terms not found in term banks from parallel/comparable web data using state-of-the-art terminology extraction and alignment methods; (4) facilities for terminology sharing and reusing within CAT tools; and (5) improvement of statistical machine translation systems through terminological data integration.

In this line, we developed EcoLexiCAT, a terminology-enhanced CAT tool that provides easy access to domain-specific terminological knowledge in context

(León-Araúz, Reimerink, & Faber, 2017; León-Araúz & Reimerink, 2018). EcoLexiCAT is freely available[2] for any user interested in translating English or Spanish environmental texts. This application integrates all features of the professional translation workflow in a stand-alone interface where a source text is interactively enriched with terminological information (i.e. definitions, translations, images, compound terms, corpus access, etc.) from different external resources: (1) EcoLexicon, a multimodal and multilingual terminological knowledge base on the environment (Faber et al., 2014; Faber et al., 2016); (2) BabelNet, an automatically constructed multilingual encyclopedic dictionary and semantic network (Navigli & Ponzetto, 2012); (3) the EcoLexicon English Corpus (EEC), powered by Sketch Engine, the well-known corpus query system (Kilgarriff et al., 2004); (4) the Inter-Active Terminology for Europe (IATE), the multilingual glossary of the EU; and (5) other external resources that can be customised by users (i.e. Wikipedia, Wordreference, Linguee, etc.).

The remainder of this chapter is organised as follows: Section 1.2 explains terminology management from the perspective of the needs and expectations of professional translators. Section 1.3 concisely describes the web-based open-source CAT Tool MateCat on which EcoLexiCAT is based, as well as the external resources used for terminology enhancement. Section 1.4 provides a detailed explanation of EcoLexiCAT and its different modules. Section 1.5 explains the procedure with which EcoLexiCAT was tested and evaluated, and discusses the results regarding users' expectations, behaviour, performance, and satisfaction. Finally, Section 1.6 presents the conclusions that can be derived from this study and outlines ideas for future research.

1.2 Translators' needs and expectations for terminology management

Any lexicographic or terminographic tool should take into account the needs of end users, in its structure and content as well as the way that the information is represented so that users can search and interact with the tool (Tarp, 2013). When translators query a resource and do not find the information needed, they lose time and their productivity goes down (search costs; Nielsen, 2008). Similarly, when translators obtain too much data (infoxication; Cornellà, 1999), which lengthens the knowledge construction time, comprehension costs (Nielsen, 2008) go up. In addition, translators do research in all phases of the translation process. Firstly during the pre-translation phase, research is carried out in order to understand the original text and its terminology. When the original message is encoded in the target text, research is performed with a view to fulfilling pragmatic requirements and searching for equivalents. Finally, in the revision phase, translators must check terminology and generally ensure the quality of their translation (Durán Muñoz, 2012: 80). Accordingly, one of the major challenges of lexicographic and terminographic resources for translators is to find the right balance between search costs and comprehension costs.

Durán Muñoz (2010, 2012) affirms that translators prefer to solve their terminological problems by consulting ready-made resources. According to her study, the most frequent resources used are (in this order): bilingual specialised dictionaries or glossaries, searches in search engines, terminological databases, monolingual specialised dictionaries, and Wikipedia (Durán Muñoz, 2012: 81). She mentions that translators do not trust the quality of multilingual resources and that searches in parallel corpora are not high on the list of preferences. When asked to classify the most frequent ISO fields (ISO 12620:1999) in the microstructure of terminological resources, translators considered the following ones to be most essential: clear and concrete definitions, equivalents, derivatives and compounds, domain specification, examples, phraseological information, definition in both languages for bilingual resources, and abbreviations and acronyms (Durán Muñoz, 2012: 82). Finally, when asked for their opinion, translators said that terminological resources should be able to do the following: (1) permit exportability and/or importability in different formats; (2) include more pragmatic information about usage and tricky translations (old usage, false friends, specific usage in a domain or region, etc.); (3) offer links to other resources to improve or increase the results; (4) improve search options; and (5) provide examples taken from real texts (idem). Quite surprisingly, although the translators in this study did not show much interest in having access to corpora, they did highlight the need for more phraseological information, pragmatic information, and examples taken from real texts. Even though this information can be extracted from corpora, translators may have been reticent to use them because it can take a long time if the right query methods are not provided.

Translation-oriented terminology management, or terminology-enhanced translation, should take into account all of these. As shown in Section 1.4, EcoLexiCAT is a tool that includes the essential fields mentioned, links to other resources, and improved search options for corpus analysis that provide the necessary pragmatic information and real text examples. All of this is available in a single-platform web-based CAT environment that has the capabilities of importing and exporting different file types and formats.

1.3 EcoLexiCAT sources

1.3.1 MateCat

EcoLexiCAT is powered by MateCat, acronym of Machine Translation Enhanced Computer Assisted Translation. It was originally a three-year research project led by a consortium composed of the international research center FBK (Trento, Italy), Translated SRL, the Université du Maine, and the University of Edinburgh. The objective was to improve the integration of machine translation (MT) and human translation (Federico et al., 2014: 129). Within the project, a computer-aided translation tool was developed, the MateCat Tool. This application is not only an industrial tool but also an open-source platform[3] that offers all of the features of a modern CAT tool, such as a text editor that divides the text to be translated in source and target segments and saves them along with their translation in a translation memory (TM).

MateCat runs as a web server and communicates with other services through open APIs. It allows communication with pre-existing TMs, terminological databases, concordance searches within the TMs and machine translation (MT) engines, from which the MT provider MyMemory (a combination of Google Translate and Microsoft Translator) is freely available. The tool has been tested in professional settings and adapted for research in MT (e.g. Bertoldi et al., 2013 as referenced by Federico et al., 2014: 131) and for educational purposes. The fact that it has an open-source version as well as its high level of flexibility made it a suitable option for the development of EcoLexiCAT. In addition, the features and operation of MateCat are basically the same as those found in most CAT tools used nowadays. Therefore, professional translators will not need to invest much time in learning how to use the tool and will benefit from the interoperability of CAT-related formats (TBX for glossaries, XLIFF for bilingual files, TMX for TMs, etc.). This enables them to use the resources generated during the translation process in other similar tools and reuse pre-existing resources (i.e. glossaries, bilingual files, and TMs) in EcoLexiCAT.

1.3.2 EcoLexicon

EcoLexicon[4] is a multilingual and multimodal terminological knowledge base on environmental science (Faber, León-Araúz, & Reimerink, 2014, 2016; San Martín et al., 2017). It is the practical application of Frame-Based Terminology (FBT; Faber et al., 2011; Faber, 2012; 2015), a theory of specialised knowledge representation that uses certain aspects of Frame Semantics (Fillmore, 1982; Fillmore & Atkins, 1992) to structure specialised domains and create non-language-specific representations. FBT focuses on conceptual organisation, the multidimensional nature of specialised knowledge units, and the extraction of semantic and syntactic information through the use of multilingual corpora.

EcoLexicon is an internally coherent information system, which is organised according to conceptual and linguistic premises at the macro- as well as the microstructural level. It currently has 4460 concepts and 23,485 terms in Spanish, English, German, French, Modern Greek, and Russian. This terminological resource is conceived for language and domain experts as well as for the general public. It targets users such as translators, technical writers, and environmental experts who need to understand specialised environmental concepts with a view to writing and/or translating specialised and semi-specialised texts.

End users interact with EcoLexicon through a visual interface with different modules that provide conceptual, linguistic, and graphical information. Instead of viewing all this information simultaneously, they can browse through the windows and select the data that is most relevant for their needs. Figure 1.1 shows the entry in EcoLexicon for FAN. When users open the application, three zones appear. The top horizontal bar gives users access to the term/concept search engine. The vertical bar on the left of the screen provides information regarding the search concept, namely its definition, term designations, associated resources, general conceptual role, and phraseology.

Each definition makes category membership explicit, reflects a concept's relations with other concepts, and specifies essential attributes and features (León-Araúz, Faber, & Montero-Martínez, 2012: 153–4). Accordingly, the definition is the linguistic codification of the relational structure shown in the concept map, at the center of the screen. Although users can configure the map to their needs, the standard representation mode (see Figure 1.1) shows a multilevel semantic network whose concepts are all linked in some way to the search concept, which is at its center.

A specialised corpus was specifically compiled for EcoLexicon in order to extract linguistic and conceptual knowledge. Currently, the corpus has over 50 million words and each of its texts has been tagged according to a set of XML-based metadata, which contain information about the language of the text, the author, date of publication, target reader, contextual domain, keywords, and so on. This was done in order to provide users with a direct and flexible way of accessing the corpus. It also allows them to constrain corpus queries based on pragmatic factors, such as contextual domains or target reader. In this way, users can compare the use of the same term in different contexts. The corpus was first made available in the Search concordances tab (center area menu just above the concept map in Figure 1.1). However, currently, the EcoLexicon English Corpus (EEC) (23 million words) is also hosted and freely available in Sketch Engine Open Corpora[5] (see Section 1.3.4). The EcoLexicon Spanish Corpus is still in the compilation phase but will be made available in the near future.

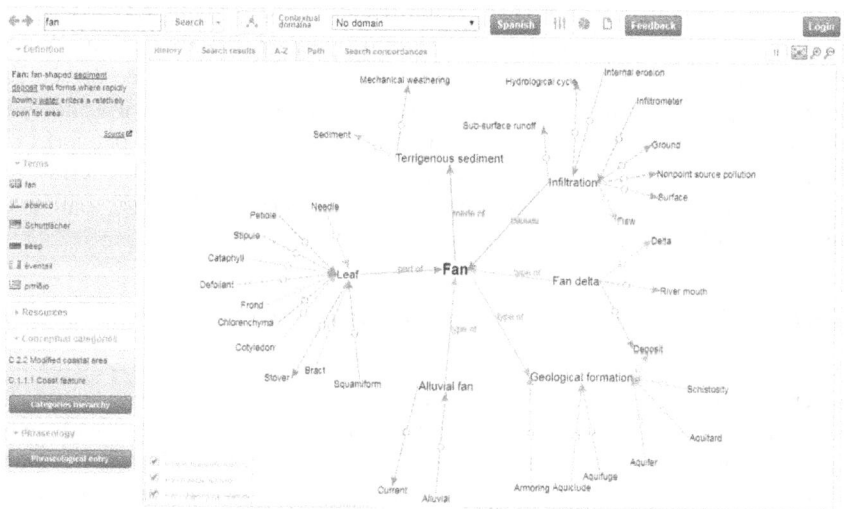

Figure 1.1 EcoLexicon user interface.

To fully exploit the contents and components of EcoLexicon for purposes of translation, we developed EcoLexiCAT. A terminological knowledge base (TKB) such as EcoLexicon provides a great amount of interconnected information in many different formats. However, in the professional translation workflow, especially when the source text has a high term density, searching in EcoLexicon, together with other resources, might cause high search and comprehension costs (see Section 1.1). EcoLexiCAT provides all this knowledge as an integral part of the translation workflow, where it is presented according to a specific context and during a specific phase of the translation process (see Section 1.4).

1.3.3 BabelNet and Babelfy

The multilingual encyclopedic dictionary and semantic network BabelNet[6] was created by linking Wikipedia to WordNet (Navigli & Ponzetto, 2012: 218). It connects concepts and named entities in a network of semantic relations, made up of about 14 million entries, called Babel synsets. Each Babel synset represents a given meaning and contains all the synonyms expressing that meaning in a range of different languages. Wikipedia and WordNet are integrated through automatic mapping and by filling in lexical gaps in resource-poor languages with MT.

BabelNet is an enormous information resource that can be accessed through an open API, and was considered to be a valuable addition to EcoLexiCAT in those cases where EcoLexicon, a manually-built resource, did not include sufficient information, for information regarding general language issues or for texts that combine environmental issues with other domains of expertise. Furthermore, the BabelNet researchers created their own algorithm, called Babelfy, for the disambiguation of polysemic words when found in the context of a particular text (Moro, Raganato, & Navigli, 2014; Moro, Cecconi, & Navigli, 2014).

Babelfy is a unified multilingual, graph-based approach to entity linking (EL; the disambiguation of named entities) and word-sense disambiguation (WSD; the disambiguation of common nouns, verbs, and adjectives). When presented with an input segment, the system extracts all the linkable fragments and lists the possible meanings of each of them according to the semantic networks of BabelNet. Evidently, this is of great help when dealing with polysemic terms. In EcoLexiCAT, the source text is disambiguated through Babelfy before matching the terms with BabelNet.

1.3.4 Sketch Engine and the EcoLexicon English Corpus

Sketch Engine (Kilgarriff et al., 2004) is an online corpus query system with a very efficient search engine and a statistical component for enhanced precision. It contains over 300 corpora in over 60 languages and allows end users to create their own corpora as well. One very interesting module is information extraction through word sketches. Word sketches are summaries of collocational information of a search term, where the term is analysed according to the verbs, modifiers, and other usual constructions that accompany it in real texts. Word sketches are created through sketch grammars that launch specific queries to a corpus. End users can create their own grammars for word sketches and thus adapt the tool to their specific needs.

With an account, users have access to pre-loaded corpora and a corpus compiler called WebBootCaT. They can download corpora, add new documents to a corpus, extract domain keywords, view texts, and generate concordances, wordlists, frequency lists, collocations, and word sketches. Sketch Engine also hosts a set of freely available open corpora that can be queried with full Sketch Engine functionalities with no need for a subscription. This made it a perfect option as a corpus query system for EcoLexiCAT, as the EcoLexicon English Corpus (EEC; León-Araúz, Reimerink, & San Martín, 2018) is hosted as an open corpus and can be accessed from external applications (i.e. EcoLexiCAT) through an API.

As previously mentioned, the EEC is tagged based on different parameters such as domain, user type, geographical variant, genre, editor, and so on, which can be used and combined to constrain queries and reduce search and comprehension costs while translating. For instance, combining the frequency functionality with that of domains or user type, users can verify if a term is more specific of one domain/user or another. Figure 1.2 is the result of searching for the verb *liquefy* and filtering the results according to user type. Not surprisingly, the verb appears more often in texts for experts than in texts addressed to the general public.

By searching the lemma *photovoltaic* and looking up its frequency according to domain, the results show that it is a term mainly linked to the domain of Renewable Energy, although it also occurs, but with much lower frequency, in Climatology and Air Quality Management (Figure 1.3).

Furthermore, the EEC was processed with the EcoLexicon Semantic Sketch Grammar (ESSG) (León-Araúz & San Martín, 2018; León-Araúz, San Martín, & Faber, 2016), a CQL-based (Corpus Query Language) (Jakubíček et al., 2010) customised sketch grammar, together with the default sketch grammar. In this way, users can benefit from default word sketches (i.e. modifiers, verbs, etc.) and semantic word sketches, as the ESSG was developed for the extraction of semantic word sketches based on some of the most common relations in terminology: generic-specific, part-whole, location, cause, and function. Thanks to

User	Frequency	Rel [%]	Items: 3 | | Total frequency: 217
P | N Expert	153	125.10	
P | N Semi-expert	52	83.60	
P | N General public	12	39.10	

Figure 1.2 Frequency of *liquefy* in the EEC according to user type.

Domain	Frequency	Rel [%]	Items: 4 | | Total frequency: 174
P | N 3.5.1 Renewable Energy	106	2,683.70	
P | N 2.7.2 Climatology	34	104.10	
P | N 3.7.5.3 Air Quality Management	17	249.40	
P | N 0 General	17	85.00	

Figure 1.3 Frequency of *photovoltaic* in the EEC according to environmental subdomain.

"microorganism" is the generic of...			"oxygen" is part of...			"tsunami" is caused by...		
		13.42			6.36			16.05
bacterium	29	10.78	atmosphere	23	9.80	earthquake	93	11.41
fungus	15	10.49	molecule	21	10.18	landslide	43	10.71
pathogen	5	9.35	compound	18	9.51	eruption	26	9.79
alga	5	8.42	water	16	8.44	water	22	8.14
virus	4	8.76	earth	15	9.26	slide	13	9.14

Figure 1.4 Semantic word sketches extracted from the EEC thanks to the ESSG.

the ESSG, users can access ready-made word sketches such as those shown in Figure 1.4, where search terms may appear related to their hyponyms (i.e. *microorganism*), the whole they are part of (i.e. *oxygen*), their underlying causes (i.e. *tsunami*), and so on.

1.3.5 IATE and other resources

IATE is the interinstitutional terminology database that has been used in the EU institutions and agencies since 2004, enhancing standardisation and promoting an official EU terminology. It has around 8.4 million terms. IATE cannot be accessed through APIs but can be downloaded. Therefore, we downloaded the set of English and Spanish terms and stored them in a database to interact with EcoLexiCAT as a fourth external resource.

Other resources generally consulted by translators have been added in a pop-up window within the interface of EcoLexiCAT. So far, these resources are *Wikipedia, Wordreference, Linguee*, the *Cambridge Dictionary*, and EcoLexicon (as deployed in a web browser). However, users can customise this window by adding new resources in the same line as the SDL Trados Studio plug-in Web Lookup or the MemoQ web search feature.

1.4 EcoLexiCAT: a terminology-enhanced translation tool

When users start a new project in EcoLexiCAT, they first access the project settings interface in Figure 1.5, where they can do the following: (1) name the project; (2) choose directionality (so far, English-Spanish or Spanish-English); (3) select a particular domain within the environment – these are in consonance with the domains according to which EcoLexicon is organised and are included in this first step as a way to classify projects and TMs for later reuse; (4) choose between general and patent segmentation rules, for the source text to be segmented accordingly; (5) optionally add an MT provider for post-editing – MyMemory is freely available, but others (e.g. Moses, DeepLingo, IP Translator) can also be added if users have an account with them; (6) optionally add users' own TMs and/or glossaries (previous or newly created for each project) – otherwise a collective TM stored in the system will be used; and (7) upload the source text. These steps, except for (3), are default options in MateCat.

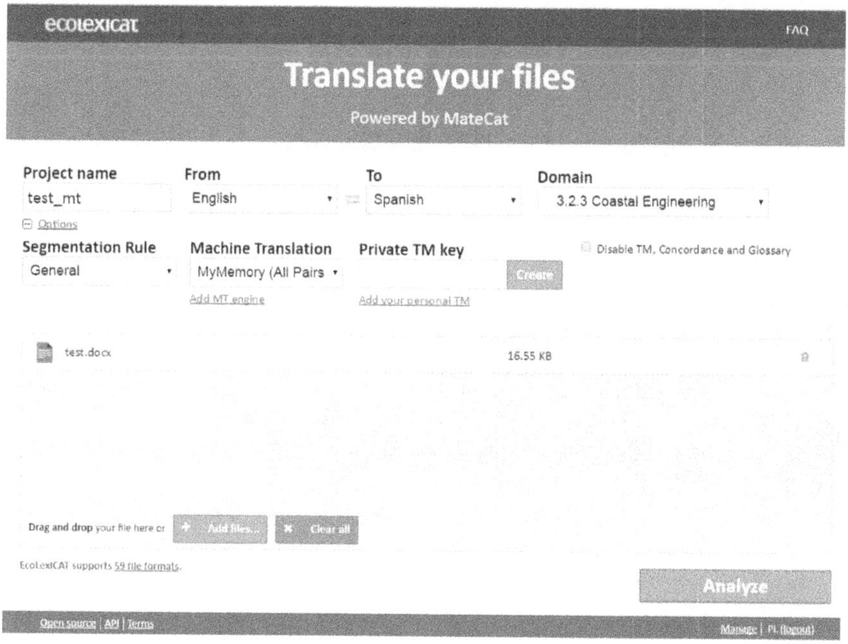

Figure 1.5 Project settings in EcoLexiCAT.

Users must register with a Gmail account in order to start using EcoLexiCAT and be able to save their work. Otherwise, they can still test the tool as an anonymous user. Users can also try a demo with previously uploaded source texts, with no need to register: a demo with MT and a demo without MT. If they register, they will need to fill in the fields shown in Figure 1.6. In Profile, users can select Student or Professional. In Area, they can choose Translation, Environmental Sciences, or Other. In the latter case they have to specify their area of expertise. Finally, after selecting their working languages, their country of origin and their country of residence, they need to specify the reason why they chose to use EcoLexiCAT: Testing, Academic Assignment, Professional Assignment, or Other. Keeping this information will help us analyse user profiles and behaviour when interacting with the tool. Moreover, it allows the classification of the resources generated (i.e. TMs), which can be used as a parallel corpus, thus enriching both the tool and the EcoLexicon Corpus.

Once the source text is uploaded, processed, and converted into a bilingual format (XLIFF), users can access the main interface (Figure 1.7), which is divided into two main sections. The left-hand section is where four external resources (i.e. EcoLexicon, BabelNet/Babelfy, Sketch Engine, and IATE) provide the terminological enhancement of the translation process. The right-hand section is where the target text is produced, an editor where the source text appears split into different segments. In the upper-right part of the editor, users may download

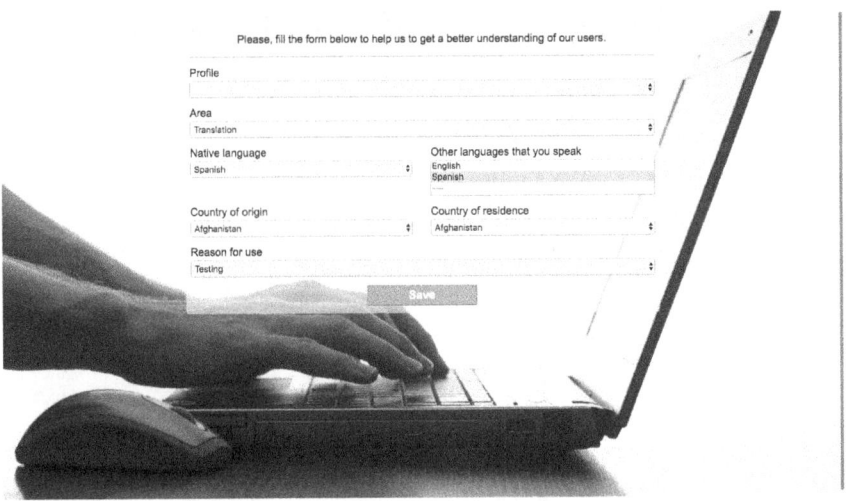

Figure 1.6 Registration form in EcoLexiCAT.

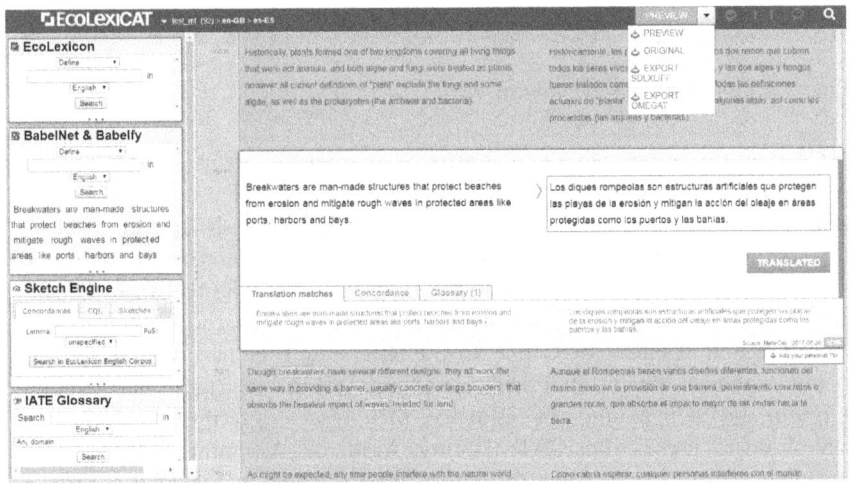

Figure 1.7 Main user's interface of EcoLexiCAT.

the target or the source text in their original format, and export the bilingual file in SDLXLIFF (SDL Trados Studio's native format) or the whole project in OmegaT's native format, another desktop open-source CAT tool. This, together with the possibility of downloading the TM and the glossary created during the project, ensures the interoperability of different formats across different CAT tools, an issue that professional translators must often deal with.

Figure 1.8 shows a segment within the editor. This editor offers the usual editing features of any CAT tool. Users can split or merge segments, copy the

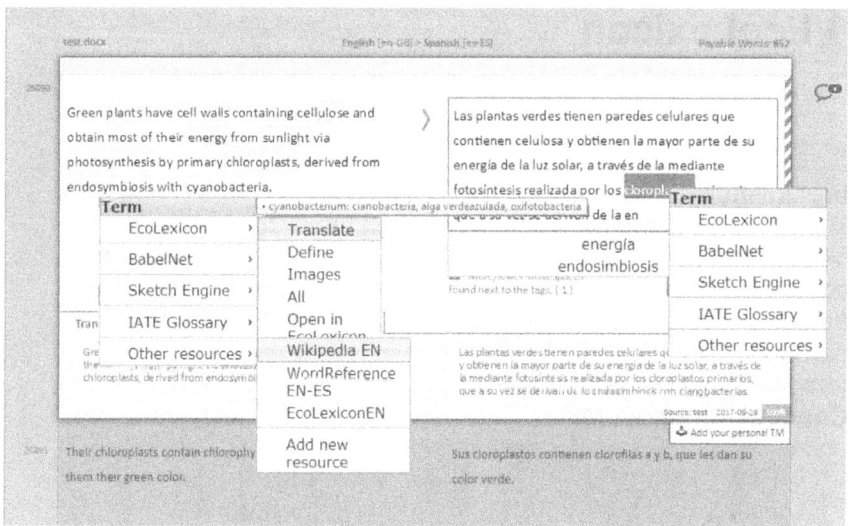

Figure 1.8 EcoLexiCAT editor.

source text in the target segment, benefit from a QA system that detects missing spaces or tags, create on-the-fly glossary entries, search for concordances within the TM, and obtain suggestions from previously stored segments in the TM or, if added, from an MT engine. Once a segment is confirmed, it is stored both in the user's TM (if it was previously uploaded or created) and in a collective TM from which other users can benefit. This, together with the fact that each task is assigned a different URL that can be shared with other translators or even reviewers, converts the tool into a collaborative environment.

However, the difference between an ordinary CAT tool and EcoLexiCAT is that the EcoLexiCAT is a terminology-enhanced translation tool. This means that the editor interacts with external terminological resources that can assist the translator during the different phases of the translation workflow. First of all, the source segment is enriched with information from EcoLexicon. This is done by lemmatising all of the words in the segment and matching them against the term entries in the TKB.

All matching terms are highlighted in yellow, and users can interact with them in three ways: (1) if they hover the mouse over them, all possible translations (equivalent terms and synonyms) are displayed in an emerging box; (2) if they click on any of them, the EcoLexicon box of the left-hand side shows both the translations and the definition; and (3) if they right-click on any of them, a scroll-down menu gives access to all the different options provided by each of the resources of the left-hand section (see Figures 1.9–1.16).

For instance, in the case of EcoLexicon, these options correspond to the data categories in the TKB that usually serve for text comprehension: translations,

◾ EcoLexicon

| Show all info of ◆ | **breakwater** | in | English ◆ |

Search

---

breakwater:
Tranlations: *dique rompeolas*, dique en talud, rompeolas
Definition: coastal defense structure, generally parallel to the coastline, made of wood, concrete or stone, to protect the coast from the impact of the wave and to provide shelter for ports and harbors.
Images
Concept *"breakwater"*

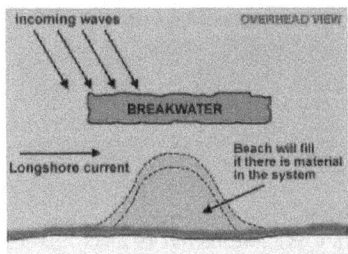 

Figure 1.9 EcoLexicon box in EcoLexiCAT.

synonyms, definitions, and images. Also from this menu, a new tab can be opened in the browser to access the EcoLexicon TKB for a more detailed analysis of the conceptual networks.

In turn, the target segment is enriched with a predictive typing feature. As soon as users start typing a word that has been matched as the translation of one of the terms in the source segment, all possible translations are shown in a drop-down list. In addition, as in the source segment, users can right-click on any term they type in the target segment and send queries to the three resources in the opposite language directionality. This is especially relevant in the case of corpus queries, as this is the resource that will usually be most useful during the text production phase.

Thus, the external resources of EcoLexiCAT interact with the segments in the editor during the different phases of the translation process, as they are terminologically enhanced for both source text comprehension and target text production tasks.

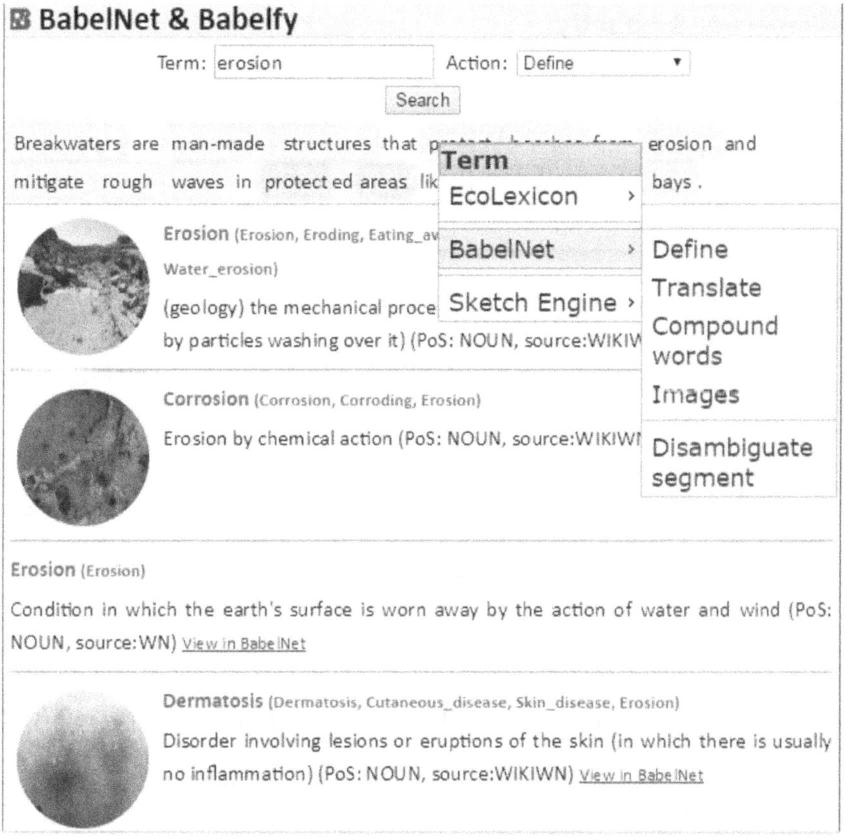

⊞ BabelNet & Babelfy

Term: [erosion]   Action: [Define ▼]

[Search]

Breakwaters are man-made structures that p~~rotect beaches from~~ erosion and
mitigate rough waves in protected areas lik ┌──────────────── bays .

Term

EcoLexicon   ›

Erosion (Erosion, Eroding, Eating_av | **BabelNet**   › | Define

Water_erosion)

(geology) the mechanical proce | **Sketch Engine** › | Translate

by particles washing over it) (PoS: NOUN, source:WIKIW | Compound words

---

Corrosion (Corrosion, Corroding, Erosion) | Images

Erosion by chemical action (PoS: NOUN, source:WIKIWN | **Disambiguate segment**

Erosion (Erosion)

Condition in which the earth's surface is worn away by the action of water and wind (PoS:
NOUN, source:WN) View in BabelNet

Dermatosis (Dermatosis, Cutaneous_disease, Skin_disease, Erosion)

Disorder involving lesions or eruptions of the skin (in which there is usually
no inflammation) (PoS: NOUN, source:WIKIWN) View in BabelNet

Figure 1.10 BabelNet and Babelfy box in EcoLexiCAT – Definitions.

In Figures 1.9–1.18, a detailed view of the external resource boxes is provided.
Figure 1.9 shows the EcoLexicon box as it appears when all features (i.e.
translations, definition, images) are requested from any of the modules where the
scroll-down menu may be activated (i.e. the EcoLexicon box itself, the BabelNet
& Babelfy box, the source segment, or the target segment). Users can also choose
to visualise these features separately.

Below the EcoLexicon box, users can find the BabelNet & Babelfy box
(Figure 1.10), where the source text is also matched against the BabelNet net
work previously disambiguated by the Babelfy algorithm. This enables the system
to propose statistically relevant candidate translations, which is a significant
advantage taking into account that BabelNet covers any specialised or general
domain and ambiguity can be frequently encountered. Furthermore, it helps the
system to arrange definitions or images in the most plausible order. For instance,
in Figure 1.10, while the first three definitions can be useful for EcoLexiCAT

Figure 1.11 BabelNet and Babelfy box in EcoLexiCAT – Translations.

users, the fourth clearly belongs to a different domain and shows a different sense of the term *erosion*.

In this box, all matched terms are highlighted in green and behave in the same manner as the terms in the source segment with regard to EcoLexicon: (1) if users hover the mouse over them, all possible translations (equivalent terms and synonyms) are displayed in an emerging box; (2) if they click on any of them, the BabelNet and Babelfy box of the left-hand side shows both the translations and the definition; and (3) if they right-click on any of them, a scroll-down menu gives access to all the different options provided by each of the resources of the left-hand section. In the case of BabelNet, these options correspond to the data categories that have been considered most interesting for translators: definitions, translations, compound words, and images. Also, from the definitions option, a new tab can be opened in the browser to access the semantic networks in BabelNet.

This box is particularly interesting for terms that are not available in EcoLexicon. This may occur when entries in EcoLexicon have not yet been included (it is a developing resource), when general language issues arise, or when the source text combines environmental terms with terms from other specialised domains. Nevertheless, users should be cautious because the Babelfy algorithm may fail or produce candidate translations that do not account for domain specificity. Being an automatically built resource based on the synsets of WordNet (a general language lexical database), BabelNet often offers a set of concepts with different levels of granularity under the same entry.

For instance, in Figure 1.11, the candidate translations go beyond equivalence, since some of the terms are hyponyms or derivatives of *erosion*. Nonetheless, when used with caution, these results can help to expand users' knowledge of the semantic network of the domain.

However, for this purpose, and especially for text production tasks, there is another option in this box, namely compound words. Figure 1.12 shows different compound terms of *erosion*, whether it acts as the head (e.g. *beach erosion*) or the modifier of the compound (e.g. *erosion control*). All of them can be clicked to access their definitions. In this way, users can browse the resource through different interconnected concepts and terms and gain a better understanding of the domain. Finally, images are the last option available from BabelNet (Figure 1.13). They can be very useful when understanding and translating complex concepts, such as processes or parts of entities, and can complement the images offered by EcoLexicon.

Below the BabelNet and Babelfy box, the Sketch Engine box appears (Figure 1.14). This box can be used to select a term from both the source and target segments and analyse its behaviour in the EcoLexicon English Corpus (as previously mentioned, the EcoLexicon Spanish Corpus is not hosted in Sketch Engine Open Corpora yet).

In the EcoLexiCAT interface, the EEC can be queried through basic concordances or CQL queries (Figure 1.14) as well as through word sketches (Figure 1.15). The output of the queries can be opened in a new tab that sends users to the website of Sketch Engine Open Corpora for a more detailed analysis. In this way, they can also use the rest of the tool's functionalities (e.g. Context, Word list, Thesaurus, Sketch Diff, etc.) and make more specific queries

🖾 BabelNet & Babelfy

Term: erosion   Action: Compound words ▾

Search

Breakwaters are man-made structures that protect beaches from erosion and mitigate rough waves in protected areas like ports, harbors and bays.

Erosion (Erosion, Eroding, Eating_away, Wearing, Wearing_away, Soil_erosion, Water_erosion)

headward erosion, beach erosion, soil erosion, erosion prediction, bank erosion, differential erosion, wind erosion, shoreline erosion, erosion control, coastal erosion, Turkish Foundation for Combating Soil Erosion, Internal erosion, lateral erosion, downward erosion

Corrosion (Corrosion, Corroding, Erosion)

stress corrosion, Corrosion Engineering, metal corrosion, crevice corrosion, corrosion resistance, corrosion inhibitors, high temperature corrosion, corrosion inhibitor, Anaerobic corrosion, corrosion prevention, electrolytic corrosion, galvanic corrosion

Figure 1.12 BabelNet and Babelfy box in EcoLexiCAT – Compound words.

⊞ BabelNet & Babelfy

Term: erosion   Action: Images   ▼

Search

Breakwaters are man-made structures that protect beaches from erosion and mitigate rough waves in protected areas like ports, harbors and bays.

Erosion (Erosion, Eroding, Eating_away, Wearing, Wearing_away, Soil_erosion, Water_erosion)

Corrosion (Corrosion, Corroding, Erosion)

Figure 1.13 BabelNet and Babelfy box in EcoLexiCAT – Images.

filtered by the features according to which the corpus is tagged (see Section 1.3.4). As previously mentioned, this information can be very useful during the text production phase (e.g. searching for modifiers or verbs that collocate with a particular noun, looking for synonyms or frequent syntactic structures, etc.). However, corpora can also help translators to understand how concepts interrelate with each other within the domain. For this reason, corpus queries are enabled from both source and target segments, as they can also be useful during the comprehension phase.

For instance, with the CQL query in Figure 1.14, users can not only access the adjectives that modify the term *breakwater* but also infer that breakwaters are usually classified according to position, material, function, etc. Furthermore, in Figure 1.15, Sketch Engine's default word sketches (e.g. modifiers and verbs) are combined with the series of customised word sketches (ESSG) especially focused on the comprehension phase, since they are based on semantic relations and thus provide knowledge rich contexts (Meyer, 2001). In Figure 1.15, the customised word sketches of the relations *is_the_generic_of* and *is_part_of* are shown for the term *mineral*. In this way, users can have quick access to part of the conceptual network of all concepts sufficiently represented in the corpus.

Figure 1.14 Sketch Engine box in EcoLexiCAT – CQL queries.

Below the Sketch Engine box appears the IATE box, which gives access to the information that the downloadable dump provides (i.e. equivalents, definitions and domains). Users can access IATE in EcoLexiCAT by typing the term they want to look up or by right-clicking the term in either the source or target segment and selecting the IATE Glossary in the scroll-down menu. In the box, they can filter the results (i.e. translations and definitions) by constraining their query according to a particular domain (e.g. politics, trade, education, environment, science, etc.) (Figure 1.16).

In the scroll down menu, the last option is Other resources, which can be accessed from both source and target segments. This is not part of the left-hand section but pops up under the open segment when users make a query in any of the default resources (i.e. Wikipedia, Wordreference, Linguee, Cambridge, EcoLexicon) or any of the ones customised by the user (Add new resource). In Figure 1.17, the Other resources box is opened through a query in Linguee. If users want to add a new resource they only need to name it and fill in its URL taking into consideration how the search term is codified in it (Figure 1.18).

☼ Sketch Engine

Concordances	CQL	Sketches

Lemma: mineral   PoS: noun ▼

Search in EcoLexicon English Corpus

mineral (noun)
freq=5252 (183.53 per million)

modifiers of "%w" 2151 40.96		nouns modified by "%w" 1984 37.78		verbs with "%w" as object 868 16.53	
clay	210 10.97	grain	84 9.45	dissolve	82 9.73
silicate	85 10.18	deposit	137 9.37	clay	13 8.87
carbonate	55 9.09	exploration	35 8.82	leach	13 8.48
sulfide	35 8.89	nutrient	39 8.79	precipitate	11 8.34
common	73 8.6	assemblage	36 8.7	extract	17 8.1
heavy	64 8.52	dust	32 8.52	rock-forming	7 8.03
valuable	33 8.5	fertilization	27 8.5	identify	39 7.98
evaporite	24 8.45	composition	48 8.39	contain	43 7.36
ore	26 8.41	resource	80 8.34	form	50 7.23
metamorphic	28 8.29	fertilizer	29 8.25	deposit	11 7.18
accessory	20 8.2	olivine	18 8.17	mine	5 7.17
oxide	22 8.16	soil	51 7.86	exploit	5 7.06
rock-forming	17 8	salt	19 7.72	compose	8 6.82
platy	16 7.91	extraction	18 7.71	concentrate	5 6.79
iron	23 7.86	nutrition	14 7.7	transform	6 6.76
rare	19 7.81	ore	15 7.66	classify	6 6.71
soluble	17 7.65	right	17 7.56	know	24 6.66
feldspar	11 7.34	nitrogen	17 7.55	remove	14 6.66
radioactive	15 7.28	replacement	13 7.5	erode	7 6.35
serpentine	10 7.23	particle	47 7.5	occur	5 6.19
hydrous	10 7.22	matter	33 7.48	find	15 5.67
secondary	18 7.18	precipitate	11 7.4	do	7 5.51
fibrous	10 7.18	aerosol	17 7.38	grow	5 5.4
magnetic	16 7.15	owner	17 7.36	carry	5 5.01
other	139 7.12	identification	13 7.34	be	136 4.99

verbs with "%w" as subject 695 13.23		"%w" is the generic of... 1125 21.42		"%w" is part of... 544 10.36	
crystallize	14 9.27	quartz	38 9.96	rock	79 10.65
melt	9 8.14	gold	26 9.4	soil	15 8.71
precipitate	6 7.99	mica	24 9.35	magma	7 8.63
dress	5 7.85	feldspar	24 9.35	melt	6 8.46
feel	6 7.61	carbonate	22 9.12	jade	6 8.46
form	19 6.9	iron	24 9.06	peridotite	5 8.18
tend	12 6.63	calcite	19 9.04	silt	5 8.13
replace	5 6.6	copper	17 8.74	crust	6 8.08
contain	16 6.57	clay	17 8.62	meteorite	6 8.07
break	7 6.57	salt	16 8.54	limestone	5 7.97
include	28 6.38	amphibole	13 8.46	planet	5 7.97
grow	6 6.19	sulfide	13 8.41	type	7 7.91
describe	6 5.56	calcium	13 8.35	earth	7 7.83
occur	14 5.37	pyroxene	11 8.28	deposit	5 7.7
remain	5 5.21	sulfur	12 8.19	material	6 7.6
become	8 5.07	zinc	11 8.19	sand	5 7.56
have	65 4.95	uranium	11 8.17	sediment	6 7.48
cause	8 4.93	magnetite	10 8.14	beach	5 7.09
be	282 4.35	coal	13 8.11	water	7 7.07
		oxide	11 7.91	group	5 7.02

Figure 1.15 Sketch Engine box in EcoLexiCAT – Word Sketches.

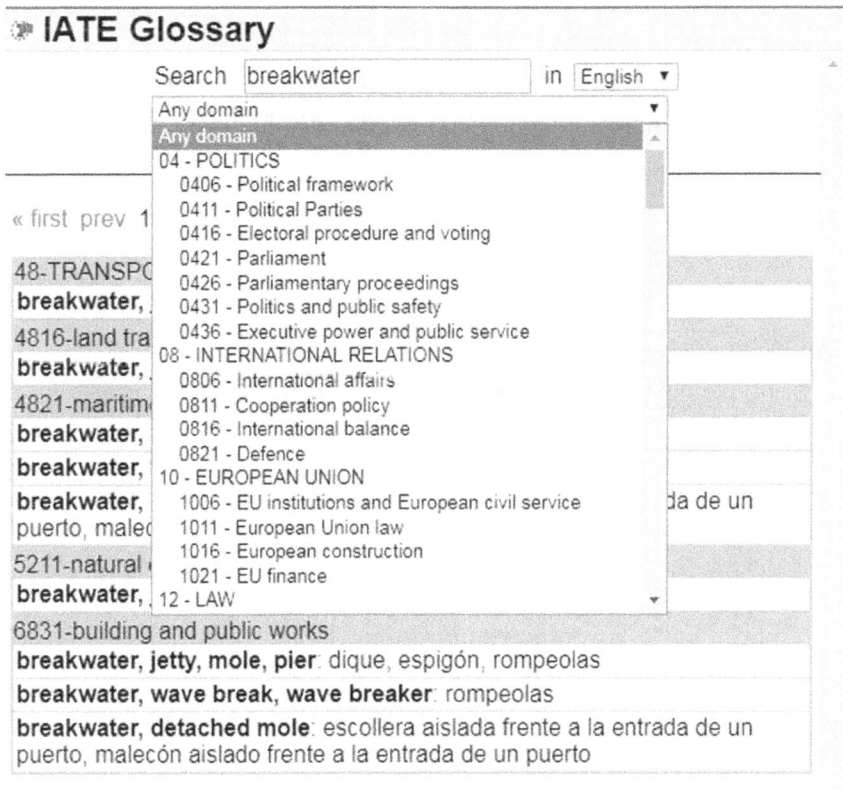

Figure 1.16 IATE Glossary in EcoLexiCAT.

Finally, there are two other features powered by MateCat that can be of interest to professional translators, as well as to lecturers and researchers. As soon as a segment is confirmed, users can open their editing log (Figure 1.19) and monitor their own performance. This includes different types of information on each segment, such as: (1) the time invested in post-editing it; (2) the suggestion source, whether it comes from MT or TMs; (3) the matching percentage between the source segment and the suggestion; and (4) the post-editing effort and tracked changes of the final target segment. These data help to raise users' awareness of their strengths and weaknesses as professional translators as well as those of the tool. For this reason, the editing log can also be exploited by translation lecturers and researchers who are interested in assessing both the work of students and/or the performance of the tool.

In this line, the revision panel (Figure 1.20) helps to perform the last phase of the translation workflow. Revisers can approve or correct all target segments. If corrected, the changes are tracked in the target cell, and revisers can use a metric

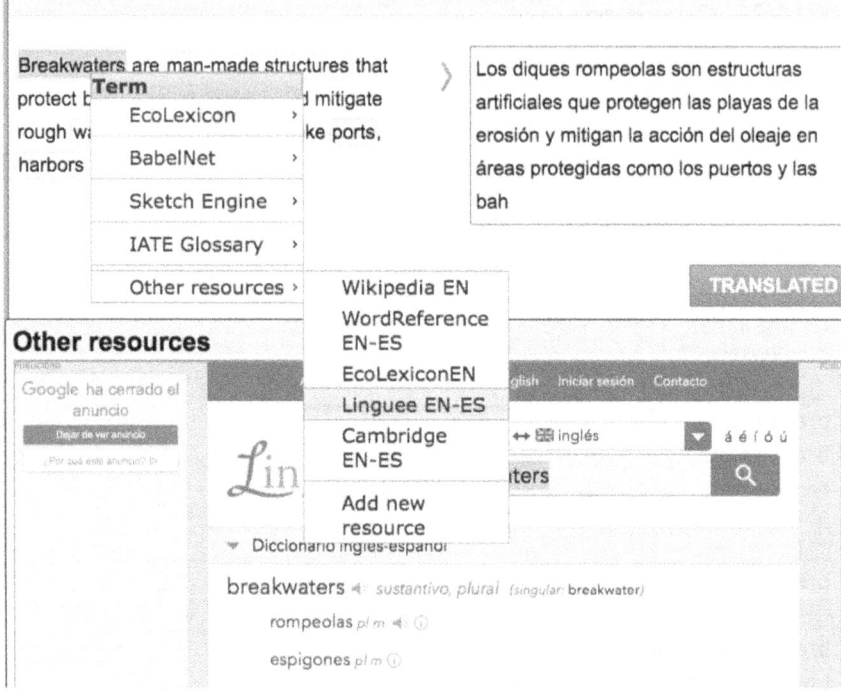

Figure 1.17 Other resources – Linguee in EcoLexiCAT.

for translation quality evaluation commonly used in the industry. This metric is based on different error types (i.e. tag issues, translation errors, terminology and translation consistency, language quality and style) and degrees (i.e. enhancement and error). At the end, users can generate a quality report that automatically scores the overall quality of the translation based on the issues highlighted by the revisers. Therefore, this feature can also be used by translation lecturers if they want to grade their students' work in a systematic way.

1.5 Evaluating EcoLexiCAT

After designing and creating EcoLexiCAT, the next logical step was to evaluate the functionalities and performance of the tool based on the experience of prospective users in order to assess whether it meets the expectations of professional translators. In the following sections the outcomes of a first study are described.

1.5.1 Experimental setup

EcoLexiCAT was evaluated by comparing the behaviour and products of two subject groups, one using EcoLexiCAT and the other one acting as a control group

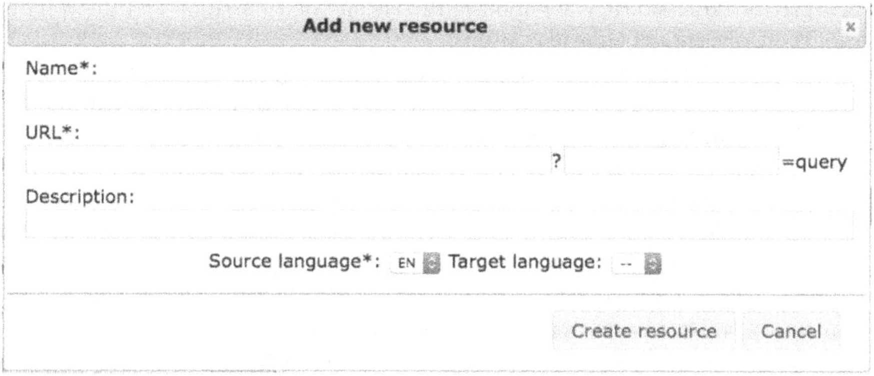

Figure 1.18 Add new resource in EcoLexiCAT.

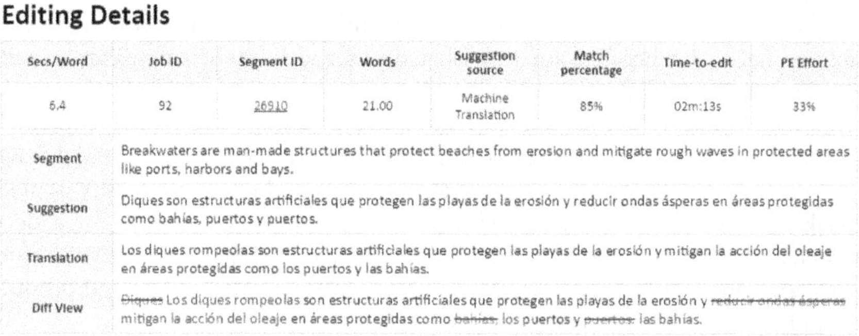

Editing Details

Secs/Word	Job ID	Segment ID	Words	Suggestion source	Match percentage	Time-to-edit	PE Effort
6.4	92	26910	21.00	Machine Translation	85%	02m:13s	33%

Segment	Breakwaters are man-made structures that protect beaches from erosion and mitigate rough waves in protected areas like ports, harbors and bays.
Suggestion	Diques son estructuras artificiales que protegen las playas de la erosión y reducir ondas ásperas en áreas protegidas como bahías, puertos y puertos.
Translation	Los diques rompeolas son estructuras artificiales que protegen las playas de la erosión y mitigan la acción del oleaje en áreas protegidas como los puertos y las bahías.
Diff View	~~Diques~~ Los diques rompeolas son estructuras artificiales que protegen las playas de la erosión y ~~reducir ondas ásperas~~ mitigan la acción del oleaje en áreas protegidas como ~~bahías,~~ los puertos y ~~puertos.~~ las bahías.

Figure 1.19 Editing log in EcoLexiCAT.

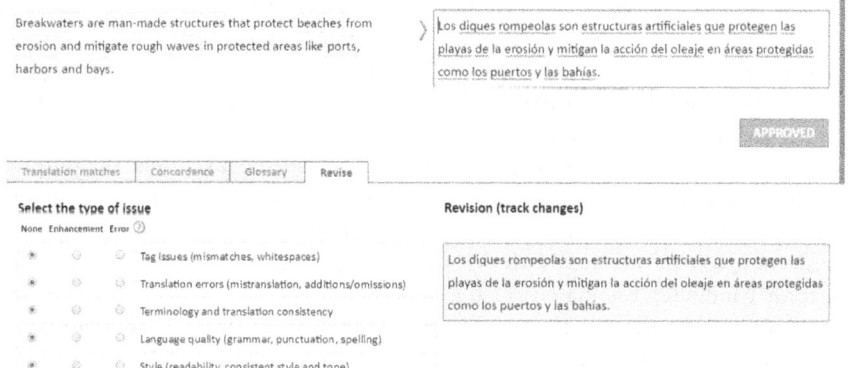

Figure 1.20 Revision in EcoLexiCAT.

and using MateCat. Both groups were made up of students from the Master's degree in Professional Translation of the Faculty of Translation and Interpreting of the University of Granada (Spain).

Prior to the translation task, participants of both groups were asked to fill out a brief questionnaire in order to collect data about their professional/training background, their expectations of terminological resources and CAT tools and their habits regarding the use of dictionaries, corpora, terminological resources, and so on when confronted with a translation assignment.

Then, the group using EcoLexiCAT was introduced to the tool, so that subjects did not waste their time getting used to its specific functionalities. Both groups were already familiar with MateCat. For this experiment, the collaborative memory option nor machine translation were used.

Afterwards, subjects were presented with a translation task consisting of two short specialised translation assignments, one English-Spanish (EN-ES) and the other Spanish-English (ES-EN), of extracts of scientific papers on the topic of Coastal Engineering, a domain widely covered in EcoLexicon. The reason for having chosen both directionalities was, firstly, to see whether behaviour and results varied according to directionality; secondly, because the only corpus available so far is the EcoLexicon English Corpus and usage examples are usually demanded during the text production phase.

Subjects were required to deliver publishable texts in two hours. Therefore, the length of each source text was under 200 words (EN-ES 194 and ES-EN 168 words). Other features of the source texts were high term density, syntactically complex sentences and collocational specificities that called for a deep understanding of both domain knowledge and written expression. Subjects were thus confronted with various challenges during the comprehension and production phases of the translation workflow.

Both groups were asked to list all the problems encountered and the resources that helped them solve each problem. The EcoLexiCAT group was allowed to use resources other than those in EcoLexiCAT only if they did not find the answer within the tool.

Finally, after finishing the assignments, EcoLexiCAT users filled out another questionnaire on the tool's usability, functionality and efficiency, three main parameters established by the ISO 9126 (2001) standard for software product evaluation. They were also asked to highlight any issues related to the functioning of the tool and propose possible improvements.

Apart from discovering the expectations of our prospective users, the purpose of this evaluation was threefold. We were able to assess user satisfaction but also user behaviour and performance. The first parameter was assessed based on the answers given by the EcoLexiCAT user group in the last questionnaire. The second parameter was based on the analysis of subjects' behaviour according to Google Analytics, whereas the third parameter was assessed by comparing the time employed and the average quality of the target texts delivered by both groups. Quality assessment was based on a scale where both translation and linguistic errors and accurate choices were accounted for. The editing log of

EcoLexiCAT and MateCat was used to see how long subjects took to translate each text.

A total of 19 students, 22 to 37 years of age, were included in the evaluation, 10 in the EcoLexiCAT group and 9 in the control group. All subjects except for one were native speakers of Spanish, 11 subjects had English as their first foreign language and 5 as their second foreign language. One subject was a native speaker of both English and Spanish and 2 did not include English as one of their official working languages during their undergraduate degree, but did have sufficient proficiency. The large majority had a translation degree (84%); the others had degrees in modern languages or related areas. Only four subjects mentioned previous professional translation experience. The different characteristics were evenly divided over both groups.

1.5.2 User expectations

In the first questionnaire, the participants were asked to classify the following features in CAT tools as essential, desirable or unnecessary: access to MT engines, access to corpora, interoperable file formats, access to terminological resources, access to terminological resources defined by users, and QA and revision options. The results in Figure 1.21 show that the most important features were found to be format interoperability, terminological resources and QA and revision options. Access to corpora was regarded as essential and desirable in the same proportion, whereas access to MT engines was deemed only desirable. This might be due to the fact that post-editing of MT is still not widely accepted by the translation community. MT engines, access to corpora, and user-defined terminological resources were the only features that were unnecessary according to a few participants.

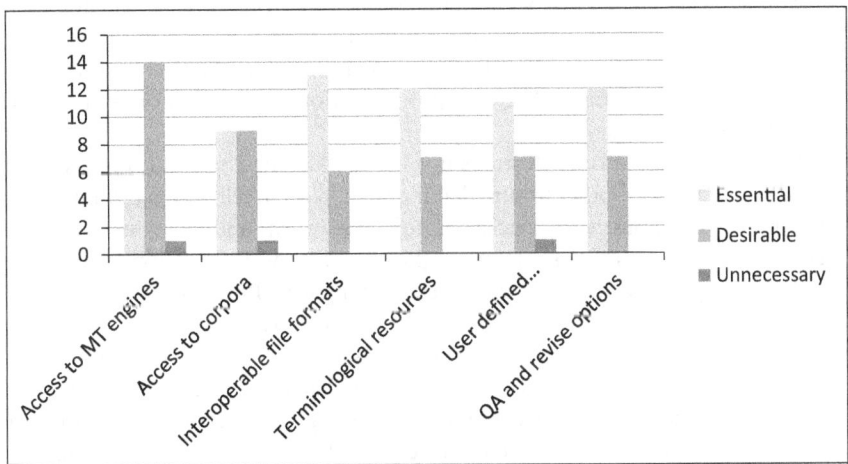

Figure 1.21 Users' expectations about CAT tools.

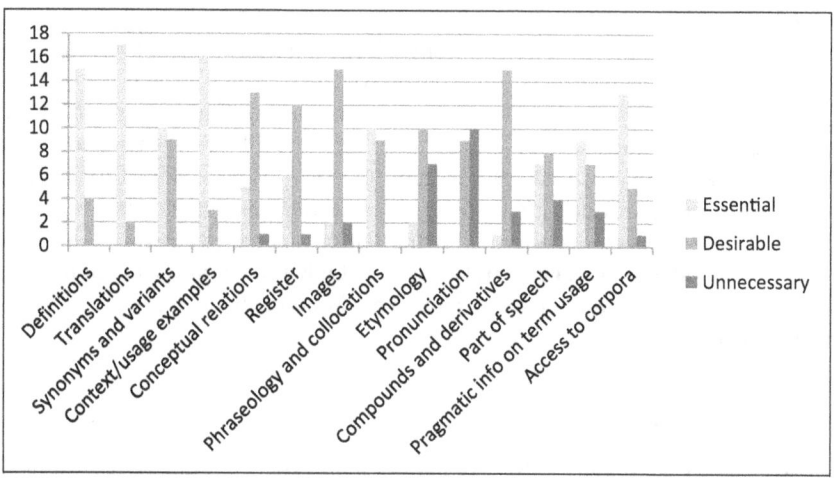

Figure 1.22 Users' expectations about terminological resources.

When asked about other features not included in the above list, most subjects could not identify any other feature that they considered to be relevant in CAT tools. Exceptions were image editors and customisable QA rules.

Participants were also asked to do the same with a set of data categories usually included in terminological resources. The data categories were: definitions, translations, synonyms and variants, context and usage examples, conceptual relations, register, images, phraseological and collocational information, etymology, pronunciation, compounds and derivatives, part of speech, pragmatic information on term usage, and access to corpora.

The results in Figure 1.22 show that definitions, translations, context and usage examples, and access to corpora are the most relevant data categories. Among desirable categories, conceptual relations, register, images and compounds and derivatives stand out. Etymology and pronunciation are the categories most often regarded as unnecessary.

When asked about other features not included in the above list, most subjects could not identify any other feature that they would deem relevant in terminological resources. Exceptions were specialised reference works and term use frequency. Among the resources that subjects use the most for their translation assignments, the following stand out: Wordreference, Linguee, Reverso Context, IATE, *Merriam-Webster, Oxford dictionaries, Collins*, esTenTen and enTenTen corpora in Sketch Engine, the BNC, CREA, the web as a corpus, Pons and Termium Plus.

The analysis of subjects' answers indicated that EcoLexiCAT meets most of users' needs and expectations, but it also highlighted certain aspects that can be improved in the tool and even EcoLexicon. For instance, currently there is a phraseology module under construction in EcoLexicon that will undoubtedly be linked to EcoLexiCAT in the future.

1.5.3 User behaviour

While completing their assignments, EcoLexiCAT subjects were monitored through Google Analytics. Prior to the evaluation task, we defined a series of "Events" according to the kind of actions that we wished to monitor. These "Events" in Google Analytics can be tracked according to a three-level structure consisting of Category (e.g. EcoLexicon), Action (e.g. definition by clicking on the terms) and Label (e.g. breakwater), which would mean that when users search for the definition of *breakwater* in EcoLexicon, by clicking in the editor, the event is stored as such. This allows us to compare the real use of each resource and the kind of queries subjects send through which kind of action (e.g. definitions, translations, images, etc. from the menu, by clicking in the editor, in the search form of each box, etc.). Table 1.1 shows a summary of the main actions tracked within each resource.

A total of 5,693 events were stored during the experiment. Most of them took place within MateCat (4,874), but among the rest of the resources, EcoLexicon stands out with 473 events (58%; Figure 1.23). EcoLexicon is followed by BabelNet, with 262 events (32%), Other resources, with 47 (6%), IATE, with 27 (3%) and Sketch Engine with 10 (1%).

From a quantitative point of view, the following Figures (1.24–1.28) show the number and type of actions performed within each of the resources, which illustrates the usefulness of both data categories of each resource (e.g. definitions, translations, images, etc.) and the way in which each category can be accessed (e.g. clicking, from the menu, writing the query, etc.). For instance, in EcoLexicon and BabelNet (Figures 1.24 and 1.25), definitions and translations are the preferred data categories and clicking is clearly the preferred action. The number of actions for definitions-click and translations-click are the same (207 and 98, respectively) because when users clicked on one of the highlighted terms in the source segment of the editor, both kinds of information were deployed in the EcoLexicon and

Table 1.1 Main actions tracked within each resource in EcoLexiCAT

Matecat	Insert-text, open-segment, delete-text, translate, search, download-original, download-translation
EcoLexicon	Definitions-click, definitions-menu, definitions-form, translations-click, translations-menu, translations-form, showAll-menu, showAll-form, images-form, images-menu, open-menu (EcoLexicon in a browser)
BabelNet	Definitions-click, definitions-menu, definitions-form, translations-click, translations-menu, translations-form, compound_words-menu, compound_words-form, images-menu, images-form
Sketch Engine	Concordance-menu, concordance-form, sketches-menu, sketches-forms, CQL-form
IATE	Search-menu, search-form
Other Resources	Load-Linguee, load-WordReference, load-Cambridge Dictionary, load-EcoLexicon

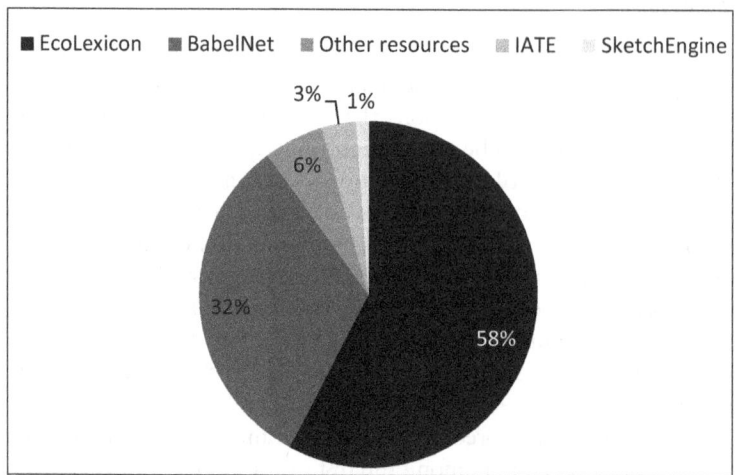

Figure 1.23 Events per resource.

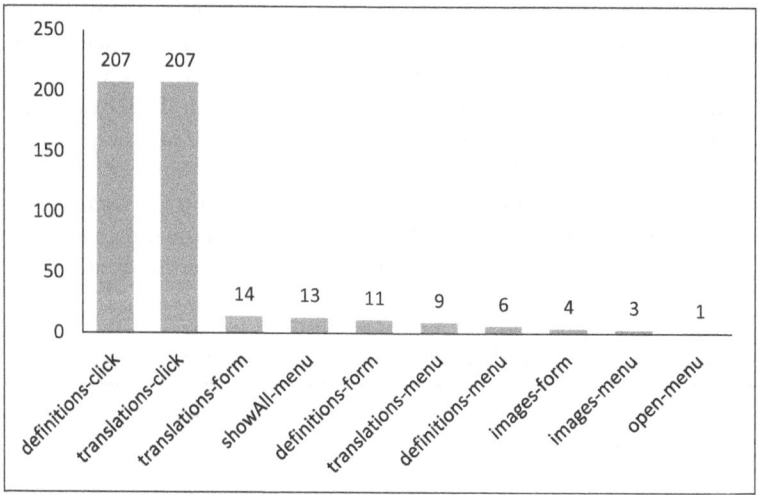

Figure 1.24 Actions performed within EcoLexicon.

BabelNet boxes at the same time. This is not surprising, as clicking is the quickest way to get to the information directly related to the content of the text.

Regarding other ways of accessing these categories, the opposite trend was observed. In EcoLexicon, subjects preferred consulting definitions and translations through the form in the box over the right-click menu access, whereas in BabelNet it was the other way round. The same happened for images.

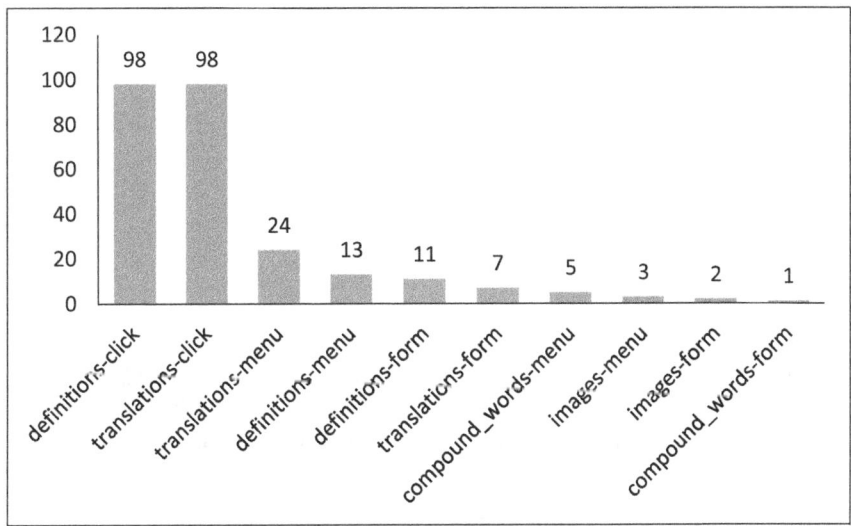

Figure 1.25 Actions performed within BabelNet.

These and compound words were the least consulted. The open-menu option was used only once. As previously mentioned, from the EcoLexicon right-click menu, users have the possibility of opening EcoLexicon in a browser for a more detailed view of the conceptual networks. Thus, adding related concepts as a new data category in the EcoLexicon box may encourage users to explore the semantics contained in EcoLexicon.

The low number of actions carried out in Sketch Engine (Figure 1.26) shows that the subjects were either not aware of the kind of information that can be extracted from a corpus or did not know how to build meaningful queries. The latter is shown by the fact that 7 of the 10 actions were simple concordance searches from the menu where only the term needs to be selected in the editor. The subjects did not seem to know the basic syntax for more complex searches that would have provided more useful information, and they do not use the more advanced functionalities of corpus analysis, such as the word sketches. However, the subjects did indicate the importance of having access to corpora in CAT tools, as most of them chose the essential or desirable options in the initial question- naire. In all likelihood, students are taught in their classes that corpus analysis is essential in the translation process, but not enough time is devoted to showing them how to actually obtain such information. In the study of Durán Muñoz (2012) mentioned earlier, professional translators did not include access to cor- pora in their preferences, which we also believe is due to lack of skills in corpus analysis and unfriendly search engines. Therefore, a user manual for EcoLexiCAT would have to provide detailed and easy-to-follow information on how to effi- ciently use the corpus options.

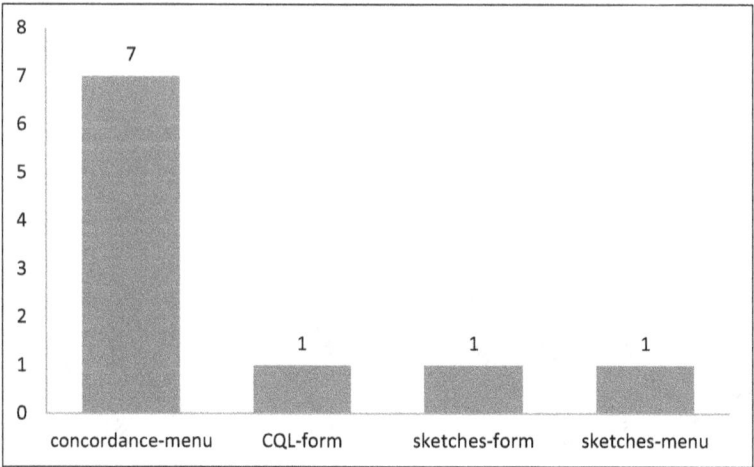

Figure 1.26 Actions performed within Sketch Engine.

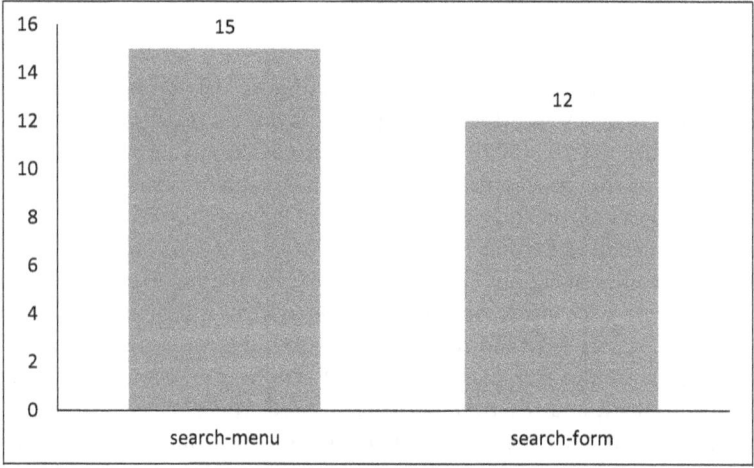

Figure 1.27 Actions performed within IATE.

In IATE (Figure 1.27), 27 actions were carried out with a slight preference for the right-click menu (15) over the use of the form in the box.

In Other resources (Figure 1.28), subjects mostly used Linguee to find translation equivalents and terms in context, and mostly during the first EN-ES translation task.

From a qualitative point of view, Table 1.2 shows the terms or text chains searched (labels) through each action within each resource. As previously stated, the definitions-click and translations-click actions are associated with the same

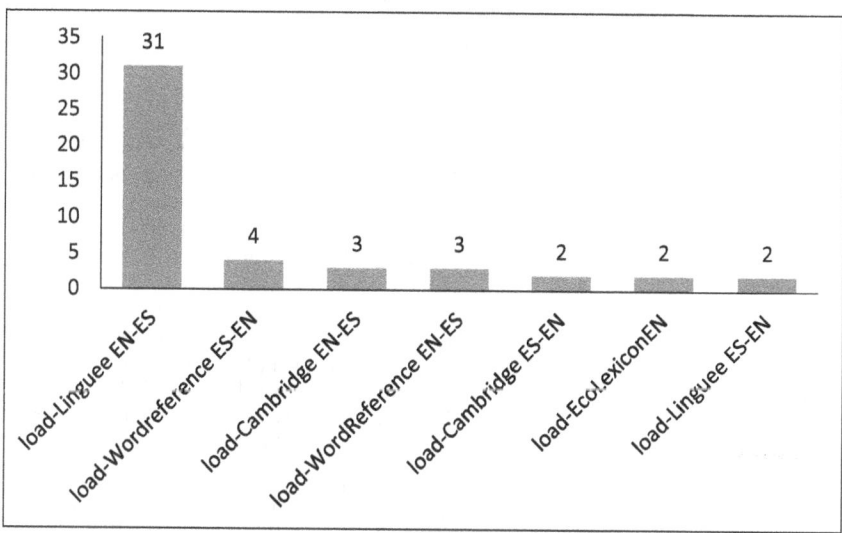

Figure 1.28 Actions performed within Other resources.

labels because they work together when users click highlighted terms in the source segment. Chains are accompanied by the number of times they were searched in brackets. Spanish chains are accompanied by their English translation in square brackets for the sake of readers' comprehension.

The data in Table 1.2 shows that there were many more searches with an English term or text chain as a starting point than Spanish. This is probably because the EN-ES task was carried out first. Both tasks were on the same subject matter and the subjects may have carried out most of the necessary research during the EN-ES task and be already familiar with many of the terms in both languages and the underlying domain knowledge.

Most searches seem to have initially looked at the options provided in EcoLexicon for the terms marked in yellow in the source text (e.g. *detached breakwaters* and *hard coastal structures*). Then when no option was given in EcoLexicon, subjects viewed the options marked in green in BabelNet, as the terms searched are clearly different at least in the searches carried out most often. The order in this process was clearly influenced by the subject matter of the tasks and the order and hierarchical structure of the terminological enhancement provided by EcoLexiCAT.

Regarding the kind of terms and chains searched for, multiword terms such as *hard coastal structure, detached breakwater*, and *artificial submerged reefs* were the ones that were most extensively researched in nearly all resources. The searched terms also matched the translation difficulties reported in the questionnaire that both subject groups filled in while translating. Almost all difficulties reported by both groups were related to the lack of previous domain knowledge, which would

Table 1.2 Terms or text chains searched through each action within each resource

Category	Action	Label
EcoLexicon	Definitions-click Translations-click	Detached breakwaters (40), hard coastal structures (29), erosion (19), hard structures (13), artificial submerged reefs (11), cliffs (10), groynes (10), coastal erosion (9), marea [tide] (8), dunes (7), oleaje [wave] (6), surge (6), coastal structures (5), sistemas [systems] (5), artificial subme rged reefs (3), beach fills (3), littoral drift (3), migración del canal [river channel migration] (3), zonas costeras [coastal zones] (3), desembocadura [river mouth] (2), estuarios [estuaries] (2), mean sea level (2), storm (2), arena [sand] (1), canal (1), prisma de marea [tidal prism] (1), sediments (1), surf zone (1), trough (1)
	Definitions-form	Dique exento [detached breakwater] (2), oleaje [wave] (2), dique [breakwater] (1), espigón [groyne] (1), hard coastal structure (1), hard structure (1), migración del canal [river channel migration] (1), soft cliff (1), storm-induced (1)
	Definitions-menu	Detached breakwaters (3), erosion (1), groynes (1), hard coastal structures (1)
	Translations-form	Oleaje [wave] (2), soft cliff (2), steady (2), sistema abierto [open system] (2), abiertos [open] (1), cierre [closure] (1), continuamente [continuously] (1), modelar [to model] (1), retool (1), river mouth (1)
	Translations-menu	Oleaje [wave] (2), surge (2), desembocadura [river mouth] (1), detached breakwaters (1), hard coastal structures (1), levels (1), soft (1)
	Images-form	Dique exento [detached breakwater] (3), espigón [groyne] (1)
	Images-menu	Detached breakwaters (1), groynes (1), hard coastal structures (1)
	Open-menu	Groyne (1)
	showAll-menu	Hard coastal structures (4), detached breakwaters (2), artificial submerged reefs (1), coastal erosion (1), coastal structures (1), erosion (1), groynes (1), hard structures (1), storm-induced (1)
BabelNet	Definitions-click Translations-click	Constante [constant] (9), sistemas abiertos [open systems] (8), bar (5), cierre [closure] (5), sistemas [systems] (4), storm (4), abiertos [open] (3), acción [action], continuamente

Table 1.2 (Cont.)

Category	Action	Label
		[continuously] (3), particular (3), remodelados [remodelled] (3), coastal (2), condiciones [conditions] (2), costeras [coastal] (2), estuarios [estuaries] (2), events (2), mareales [tidal] (2), nourishments (2), oleaje [wave] (2), rápida [rapid] (2), reefs (2), shoreface (2), sujetos [subjects] (2), waning (2), apex (1), artificial (1), barred (1), described (1), design (1), detached (1), erosion (1), events (1), evolución [evolution] (1), groynes (1), hard (1), influencia [influence] (1), marea [tide] (1), migración [migration] (1), sediment transports (1), sediments (1), submerged (1), trough (1), wave flume (1), zonas [zones] (1)
	Definitions-form	Coastal structure (1), storm-induced (1), banco [bank](1), banco de arena [sandbank] (1), detached breakwater (1), dique exento [detached breakwater] (1), erosión costera [coastal erosion] (1), erosión litoral [sea erosion] (1), escollera [rubble-mound] (1)
	Definitions-menu	Groynes (4), nourishments (2), events (2), shoreface (1), significantly (1), storm (1), storm-induced (1)
	Translations-form	Sistemas abiertos [open systems] (2), bar (1), cierre [EN] (1), cierre [ES] [closure] (1), escollera [rubble-mound] (1), steady (1)
	Compound words-menu	Cliff (2), storm (2), storm-induced (1)
	Images-menu	Erosion (1), events (1), groynes (1)
	Images-form	Dique exento [detached breakwater] (2)
	Compound words-form	Bar (1)
Sketch Engine	Concordance-menu	Groynes (3), nourishment (1), shoreface nourishment (1), shoreface nourishments (1), storm-induced (1)
	CQL-form	[search:[lemma="closure"][lemma="condition"], default_attribute:word] (1)
	Sketches-form	Nourishment (1)
	Sketches-menu	Groyne (1)
IATE	Search-menu	[lan:ENnone] (3), groynes (3), erosion (2), groyne (2), storm-induced (2), abiertos [open] (1), sistema [system] (1), surge (1)
	Search-form	Abiertos [open] (1), breakwater (1), detached breakwater (1), detached breakwaters (1), dique exento [detached breakwater] (1), erosion (1), espigón [groyne] (1), exento [detached] (1), hard structure (1), soft cliff (1), soft lift (1), surge (1)

(*continued*)

Table 1.2 (Cont.)

Category	Action	Label
Other Resources	Load-Linguee EN-ES	Storm-induced (5), detached breakwaters (4), soft cliffs (4), hard coastal structures (2), storm-induced erosion (2), against erosion (1), are built to (1), artificial submerged reef (1), coastal erosion (1), coastal structures (1), Design (1), no remedy (1), remedy (1), sandy dunes (1), significantly (1), surge levels (1)
	Load-WordReference ES-EN	Cierre [closure] 1), en particular [particularly] (1), marea [tide] (1), remodelados [remodeled] (1)
	Load-Cambridge EN-ES	Estuary (1), soft (1), storm-induced (1)
	Load-WordReference EN-ES	Detached breakwater (1), estuary (1), storm-induced (1)
	Load-Cambridge ES-EN	Constante [constant] (1), remodelados [remodeled] (1)
	Load-EcoLexiconEN	Soft cliffs (1), storm-induced (1)
	Load-Linguee ES-EN	Están sujetos [are subject to] (2)

impair the understanding of certain concepts, and to the lack of equivalences in the resources they checked. Most of the resources that helped them solve their difficulties were the ones included in EcoLexiCAT, with the exception of some general language dictionaries and parallel texts found in the web. A few students also reported phraseological issues, which explains the queries of chains like *storm-induced*, *system*, or *subject to*.

It may seem that some terms that were searched for would not cause any difficulty for their translation, such as *erosion* (19) and *cliffs* (10) in EcoLexicon. However, when working in a subject domain for the first time, researching more general terms and finding out how these concepts are related to others often helps to construct an initial mental representation of the domain.

What does seem strange is that some students looked for general language expressions, such as *continuamente* [continuously] and *significantly* in specialised resources such as EcoLexicon or BabelNet instead of using the Other resources menu. This indicates that maybe these resources should also be included in the left-hand side of the screen as a fifth box instead of as a pop-up window. Apart from that, some subjects used the definitions box in EcoLexicon (action: definitions-form) to find terms already marked in yellow in the text editor, such as *coastal structure*. The cause for this apparently strange behaviour can be explained by the fact that the subjects in our study were students with hardly any professional experience, although many had a previous translation degree and were students of a master's degree in translation.

1.5.4 User performance

All target texts were evaluated by one reviser to ensure that the same criteria were applied in all cases. To assess the quality of the target texts of both groups, ten translation problems were identified for both the English-Spanish and the Spanish-English assignment. The problems identified were based on those that the subjects mentioned repeatedly and on the reviser's expertise in the text type and domain. Depending on how well the subjects solved these problems, they could obtain up to 10 translation points. On the other hand, the language errors in both Spanish and English were deducted from a maximum grade of 10. The final grade was then the average between the translation points obtained and the linguistic quality of the target text (see Table 1.3).

For example, one translation problem of the English-Spanish assignment was finding the correct terminological equivalent in Spanish for the different types of current (longshore, tidal and rip current). Another problem was understanding the exact location of a groyne in "perpendicular or slightly oblique to the shoreline extending into the surf zone (generally slightly beyond the low water line)." An example of a translation problem in the Spanish-English assignment was understanding that *bocana* and *desembocadura* are synonyms and can both be translated as *river mouth*.

The EcoLexiCAT group outperformed the control group in both directionalities, although only slightly in the ES-EN assignment. The average quality of the target texts of both groups was not very high. This is understandable because most subjects of both groups did not have any professional translation experience or previous knowledge on the environmental domain. The results are promising, though, as EcoLexiCAT helps to obtain a better target text in less time.

It is also interesting that the control group used very similar resources to solve the translation problems to those included in EcoLexiCAT: EcoLexicon, BabelNet, Wordreference, IATE, Linguee, and Wikipedia.

Table 1.3 User performance: quality

	EcoLexiCAT EN-ES	Control EN-ES	EcoLexiCAT ES-EN	Control ES-EN
	6.8	6.1	6.9	7.1
	4.5	3.9	7.5	6
	8.4	7.6	8.1	6.6
	7.4	3.9	6.4	5
	8.0	4	7.1	4.6
	7.8	4.8	5.4	6.5
	7.4	6.3	6.1	7.3
	6.5	6.5	4.5	6.6
	7.6	6.1	7.6	6.1
	5.3		4.5	
Average	6.9	5.5	6.4	6.2

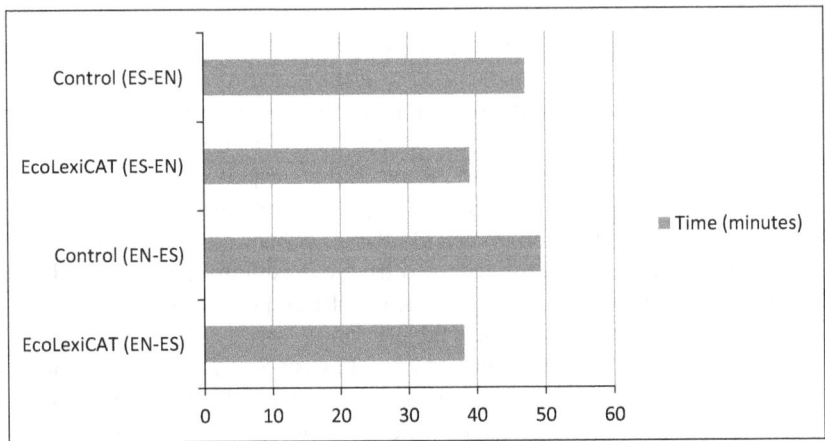

Figure 1.29 User performance: time invested.

In terms of the time invested (Figure 1.29), in both directionalities the EcoLexiCAT group outperformed the control group. Surprisingly, the EcoLexiCAT group took longer in the ES-EN assignment than in the EN-ES one, whereas the control group took longer in the EN-ES assignment, which is striking because even though it was a shorter source text, the assignment involved translating into a non-mother tongue of most of the subjects.

1.5.5 User satisfaction

Generally speaking, the subjects belonging to the EcoLexiCAT group believed that the tool was very useful (60%) or useful (40%). No subjects answered "not very useful" or "useless" when asked about the general usefulness of the tool for the translation of environmental texts.

The parameters of functionality, usability and efficiency were evaluated based on the rating of different items in a 1-to-5 scale, where 1 was the lowest and 5 the highest rate. After that, subjects could fill a free-text field to report problems, make suggestions for improvement or highlight the tool's strengths.

Regarding functionality (Figure 1.30), the subjects were asked whether the tool contained suitable features for: (1) the translation of environmental texts (80% answered 4 and 20% 5); (2) the comprehension phase of an environmental text (80% answered 4 and 20% 3); and (3) the production phase of an environmental text (50% answered 4, 40% 3, and 10% 2). This implies that EcoLexiCAT is more comprehension-oriented and that future improvements should head for increasing assistance in production-oriented tasks.

After that, they were consulted on the type of information provided and the usefulness of external resources. They were asked whether the information provided was: (1) reliable and precise (50% answered 4, 30% 5 and 20% 3); and

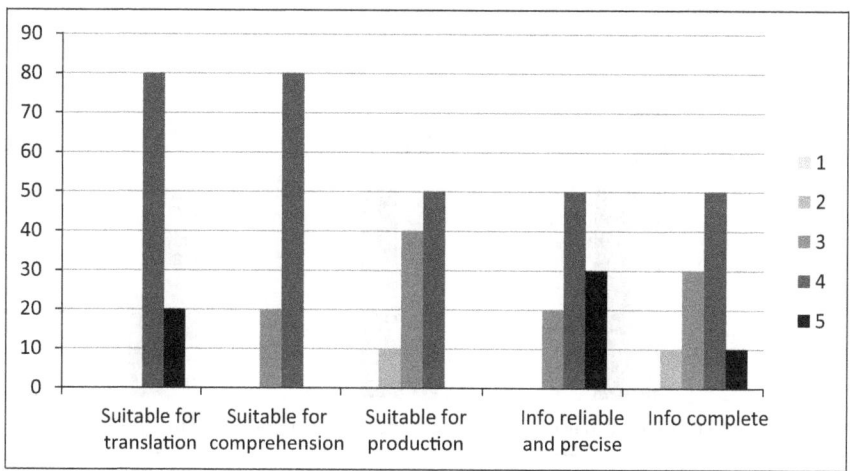

Figure 1.30 Functionality of EcoLexiCAT.

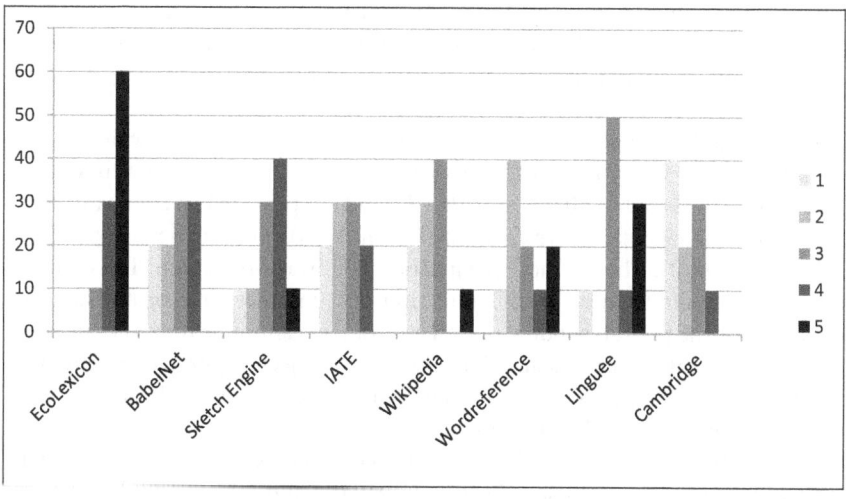

Figure 1.31 Usefulness of external resources.

(2) complete (50% answered 4, 30% 3, 10% 2, and 5). These figures call for a continuous extension, improvement and maintenance of terminological resources. One of the subjects stated that the only improvement that could be done to the tool was to extend the knowledge contained in EcoLexicon. In this line, a few subjects reported as a translation difficulty the fact that in all resources synonyms and term variants are listed with no clues on how to choose one or another.

When asked to rate the usefulness of external resources during their assignments (Figure 1.31), EcoLexicon, Sketch Engine, Linguee, and Wikipedia were rated

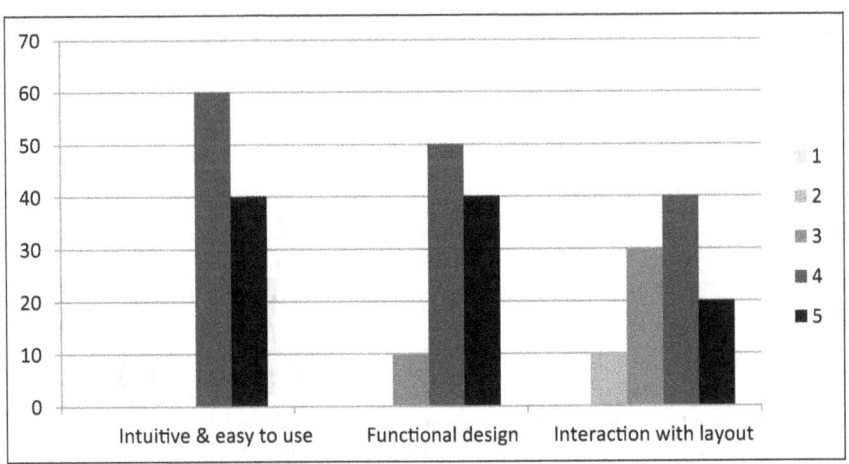

Figure 1.32 Usability of EcoLexiCAT.

best. However, this is not exactly in consonance with users' behaviour, since especially Sketch Engine was hardly consulted and Wikipedia was not consulted at all. This shows how user introspection cannot be the only method to evaluate a tool.

As for usability (Figure 1.32), subjects were asked whether EcoLexiCAT: (1) was intuitive and easy to use (60% answered 4 and 40% 5); (2) had a functional design (50% answered 4, 40% 5, and 10% 3); and (3) provided an adequate interaction with the layout (e.g. resizing of the windows) (40% answered 4; 30% 3, 20% 5, and 10% 2). The interaction with the layout was worst rated. Thus, future improvements should head in this direction. For instance, one of the subjects reported that the pop-up window of Other resources cannot be resized or moved, hindering a confortable visualisation. Another subject suggested adding the Other resources window to the left-hand side of the screen, as already inferred from the analysis of users' behaviour.

Finally, efficiency (Figure 1.33) was assessed based on whether the information was loaded at the right speed (60% answered 4 and 40% 5) and fluency: (1) user interaction with the editor (50% answered 4, 20% 5, 20% 3, and 10% 2); (2) interaction of the editor with external resources (50% answered 4, 30% 3, and 20% 5); (3) user interaction with external resources (50% answered 3, 40% 4, and 10% 5).

Some subjects reported several bugs and efficiency issues regarding the predictive typing feature in the target segment, which did not work well in the case of multiword terms. Also, certain plural multiword terms in Spanish were not lemmatised properly and thus not recognised as terms in EcoLexicon. Nonetheless, several subjects pointed out that the quick access to so many resources in the same interface, as well as the fact that the search terms do not

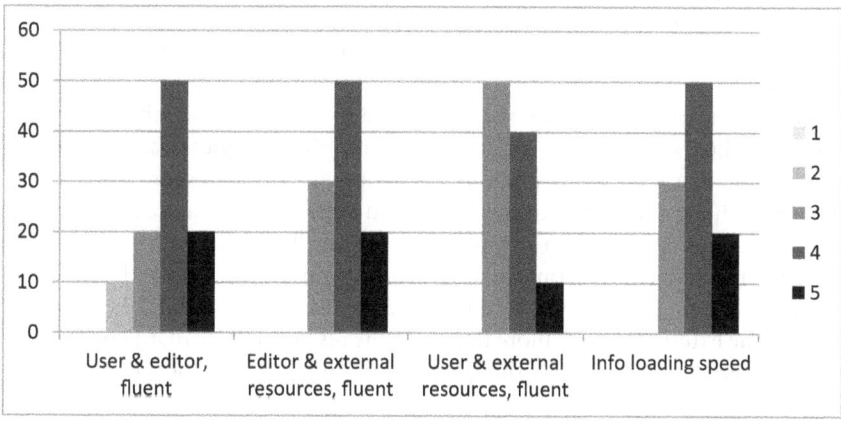

Figure 1.33 Efficiency of EcoLexiCAT.

need to be typed, is the main strength of the tool, which is the whole idea behind our concept of terminology enhancement.

The three parameters point to a favourable evaluation of EcoLexiCAT, although efficiency will be the first aspect to be improved in the future. Although it is difficult to generalise the results obtained from this first experiment, they definitely indicate that integrating terminology management in the translation workflow in a stand-alone interface improves the quality of the translation and reduces the time spent on the task.

1.6 Conclusions and future work

In this paper we have presented EcoLexiCAT, a terminology-enhanced tool that enriches both source and target segments with terminological information from various external resources in an interactive environment. The tool has been designed to meet the expectations of professional translators regarding terminology management and it seems to be on the right path according to the first evaluation study. However, it still needs to be assessed by more prospective users. More extensive studies with larger samples, including professional translators and the machine translation option, will be carried out in the future. The larger the sample, the more data we will be able to collect on how the tool can be improved based on how translators behave. For instance, if EcoLexiCAT were extensively used, we would be able to know which terms/concepts are more susceptible to being researched through a definition or an image.

New external resources will be added. However, some of the resources most reported by the subjects cannot be included for technical reasons, since entries do not have different URLs. Furthermore, EcoLexicon is currently being linked to other encyclopedic (i.e. DBpedia) and environmental resources (i.e. GEMET,

AGROVOC) by means of Linked Data (León-Araúz et al., 2011). Once the TKB is fully integrated into the Linguistic Linked Open Data Cloud, EcoLexiCAT will also benefit from reliably disambiguated encyclopedic and specialised term entries.

Two other features from EcoLexicon will also be added once they are ready. These are the EcoLexicon Spanish Corpus and phraseological patterns from a new module that is currently under construction as well as more extensive information on the distinction of term variants and synonyms. Semantic relations will also be added, since the open-menu option was hardly used.

Apart from benefiting from the improvements in EcoLexicon and its corpus, EcoLexiCAT will also be improved to make the interaction among the user, the tool and the external sources more fluent. As the results indicate that EcoLexiCAT was considered more useful for the comprehension phase than for the production phase, we plan to add other sources to improve on the latter.

Acknowledgements

This research was carried out as part of the project FFI2017-89127-P, Translation-oriented Terminology Tools for Environmental Texts (TOTEM), funded by the Spanish Ministry of Economy and Competitiveness. The authors would like to thank the master students who participated in the evaluation of EcoLexiCAT.

Notes

1　http://taas-project.eu/
2　Temporarily hosted in manila.ugr.es:9966
3　www.matecat.com/open-source
4　https://ecolexicon.ugr.es
5　https://the.sketchengine.co.uk/open
6　babelnet.org

References

Bertoldi, N., Cettolo, M., & Federico, M. (2013). Cache-based online adaptation for machine translation enhanced computer assisted translation. In Proceedings of the MT Summit XIV, Nice, France, September, pp. 35–42.

Cornellà, A. (1999). A mayor desarrollo informacional, menor infoxicacion. El profesional de la información. Available at: www.elprofesionaldelainformacion.com/contenidos/1999/septiembre/a_ mayor_desarrollo_informacional_menor_infoxicacion.html.

Durán Muñoz, I. (2010). Specialized lexicographical resources: A survey of translators' needs. In S. Granger & M. Paquot (eds.), *eLexicography in the 21st century: New challenges, new applications. Proceedings of ELEX2009. Cahiers du Cental.* Vol. 7. Louvain-La-Neuve: Presses Universitaires de Louvain, pp. 55–66.

Durán Muñoz, I. (2012). Meeting translators' needs: Translation-oriented terminological management and applications. *Journal of Specialised Translation*, 18, pp. 77–92.

Faber, P., León-Araúz, P., & Reimerink, A. (2011). Knowledge representation in EcoLexicon. In N. Talaván, E. Martín Monje, & F. Palazón (eds.), *Technological innovation in the teaching and processing of LSPs. Proceedings of TISLID, 10.* Madrid: Universidad Nacional de Educación a Distancia, pp 367–85.

Faber, P. (ed.) (2012). *A cognitive linguistics view of terminology and specialized language.* Berlin: Mouton de Gruyter.

Faber, P. (2015) Frames as a framework for terminology. In H. J. Kockaert & F. Steurs (eds.), *Handbook of terminology,* 1. Amsterdam: John Benjamins Publishing, pp. 14–33.

Faber, P., León-Araúz, P., & Reimerink, A. (2014). Representing environmental knowledge in EcoLexicon. In *Languages for specific purposes in the digital era. Educational linguistics,* 19. Cham: Springer, pp. 267–301.

Faber, P., León-Araúz, P., & Reimerink, A. (2016) EcoLexicon: New features and challenges. In I. Kernerman, I. Kosem Trojina, S. Krek, & L. Trap-Jensen (eds.), Proceedings of *GLOBALEX 2016: Lexicographic Resources for Human Language Technology in conjunction with the 10th edition of the Language Resources and Evaluation Conference,* Portorož, pp. 73–80.

Federico, M. et al. (2014). The MateCat Tool. *In Proceedings of COLING 2014, the 25th International Conference on Computational Linguistics: System Demonstrations,* Dublin, Ireland, August 23–29, pp. 129–32.

Fillmore, C. J. (1982). Frame semantics. In The Linguistic Society of Korea (ed.), *Linguistics in the morning calm.* Seoul: Hanshin, pp. 111–37.

Fillmore, C. J., & Atkins, B. T. S. (1992). Toward a frame-based lexicon: The semantics of RISK and its neighbors. In A. Lehrer & E. Kittay (eds.), *Frames, fields and contrasts: New essays in semantic and lexical organization.* Hillsdale, NJ: Erlbaum, pp. 75–102.

Gornostay, T. (2014). Dreams of better terminology tools. *Multilingual Magazine* April/May, pp. 44–5.

International Organization for Standarization (ISO). (1999). *ISO 12620: Computer applications in terminology – Data categories.* Geneva: ISO.

International Organization for Standarization (ISO). (2001). *ISO 9126: Software quality characteristics.* Geneva: ISO.

Kilgarriff, A, Rychly, P., Smrz, P., & Tugwell, D. (2004). The sketch engine. In *Proceedings of the 11th EURALEX International Congress.* Lorient: EURALEX, pp. 105–16.

León-Araúz, P., Reimerink, A., & Faber, P. (2017) EcoLexiCAT: A terminology-enhanced translation tool for texts on the environment. In I. Kosem, J. Kallas, C. Tiberius, S. Krek, M. Jakubíček, & V. Baisa, (eds.), *Electronic lexicography in the 21st century. Proceedings of eLex 2017 Conference.* Brno: Lexical Computing, pp. 321–41.

León-Araúz, P., San Martín, A., & Faber, P. (2016) Pattern-based Word Sketches for the Extraction of Semantic Relations. In *Proceedings of the 5th International Workshop on Computational Terminology (Computerm2016).* Osaka, Japan: Coling, pp. 73–82

León-Araúz, P., & Reimerink, A. (2018) Evaluating EcoLexiCAT: A Terminology-Enhanced CAT Tool. In *Proceedings of the Eleventh International Conference on Language Resources and Evaluation (LREC 2018),* edited by Calzolari, N., Choukri, K., Cieri, C., Declerck, T., Goggi, S., Hasida, K., Isahara, H., Maegaard, B., Mariani, J., Mazo, H., Moreno, A., Odijk, J., Piperidis, S., & Tokunaga, T. Paris, France: European Language Resources Association (ELRA).

León-Araúz, P., San Martín, A., & Reimerink, A. (2018) The EcoLexicon English Corpus as an open corpus in Sketch Engine. In *Proceedings of the 18th EURALEX International Congress,* edited by Čibej, J., Gorjanc, V., Kosem, I., & Krek, S., pages 893–901. Ljubljana: Euralex.

León-Araúz, P., & San Martín, A. (2018) The EcoLexicon Semantic Sketch Grammar: From knowledge patterns to word sketches. In *Proceedings of the LREC 2018 Workshop "Globalex 2018 – Lexicography & WordNets,"* edited by Kerneman, I., & Krek, S., pages 94–99. Miyazaki: Globalex.

León Araúz, P., Faber, P., & Montero Martínez, S. (2012). Specialized language semantics. In P. Faber (ed.) *A cognitive linguistics view of terminology and specialized language, 20.* Berlin, Boston: De Gruyter Mouton, pp. 95–175.

León Araúz, P., Faber, P., & Magaña Redondo, P. J. (2011) Linking domain-specific knowledge to encyclopedic knowledge: An initial approach to linked data. In *Proceedings of the 2nd Workshop on the Multilingual Semantic Web (The 10th International Semantic Web Conference)*, Bonn, Germany, 23 October 2011, pp. 68–73.

Meyer, I. (2001). Extracting knowledge-rich contexts for terminography: A conceptual and methodological framework. In D. Bourigault, C. Jacquemin, M. C. L'Homme (eds.), *Recent Advances in Computational Terminology*, pp. 279–302.

Moro, A, Raganato, A., Navigli, R. (2014). Entity Linking meets Word Sense Disambiguation: A unified approach. *Transactions of the Association for Computational Linguistics (TACL)*, 2, pp. 231–244.

Moro, A., Cecconi, F., & Navigli, R. (2014). Multilingual Word Sense Disambiguation and entity linking for verybody (2014). Proc. of the 13th International Semantic Web Conference, Posters and Demonstrations (ISWC 2014), Riva del Garda, Italy, 19–23 October 2014, pp. 25–28.

Navigli, R., & Ponzetto, S. P. (2012). BabelNet: The automatic construction, evaluation and application of a wide-coverage multilingual semantic network. *Artificial Intelligence*, 193, pp. 217–250.

Nielsen, S. (2008). The effect of lexicographical information costs on dictionary making and use. *Lexikos*, 18, pp. 170–189.

Tarp, S. (2013). What should we demand from an online dictionary for specialized translation? *Lexicographica – International Annual for Lexicography*, 29(1), pp. 146–162.

Tudhope D., Koch T., & Heery R. (2006). Terminology Services and Technology: JISC state of the art review. Available at: www.ukoln.ac.uk/terminology/JISC-review2006.html.

2 A corpus study of sustainability translation and communication in China using multilingual environmental terminologies

Meng Ji, Stefan Jensen, Jiajin Xu, and Yunlong Jia

2.1 Development of the GEneral Multilingual Environmental Thesaurus (GEMET)

The GEneral Multilingual Environmental Thesaurus (GEMET)[1] has been developed as an indexing, retrieval and control tool for the European Environment Agency (EEA), Copenhagen, its network of Member Countries (Eionet) and other environmental stakeholders since the beginning of 1996 (Budin, 2007). During the following five years, development work was rather intense and has been undertaken by partners of the EEA in Germany, Austria, Italy, Spain and Sweden as well as with the Environment Program of the United Nations (UNEP). The basic idea for the development of GEMET was to use the best of the available multilingual thesauri as a starting point, in order to save time, energy and funds. GEMET was conceived as a "general" thesaurus, aimed to define a common general language, a core of general terminology for the environment. Specific thesauri and descriptor systems (e.g. on Nature Conservation, on Wastes, on Energy, etc.) have been excluded from the first step of development of the thesaurus and have been taken into account only for their structure and upper-level terminology. A set of existing national – partially multilingual – thesauri from Germany, Italy, The Netherlands, Spain and France as well as vocabulary from EEA assessments and from the EU Commissions Eurovoc thesaurus[2] were merged (Steinberger, Pouliquen and Hagman, 2002).

The merging has been performed both on conceptual and formal basis. Coinciding concepts in the different thesauri have been identified and scored. Like in other multilingual thesauri, for example, UNEP's Infoterra EnVoc, a neutral alphanumerical notation allows the identification of a concept independently on the user's language. The resulting 6,562 terms have been arranged in a classification scheme made of 3 super-groups, 30 groups plus 5 accessory, instrumental groups. Each descriptor has been arranged in a hierarchical structure headed by a Top Term. The level of poly-hierarchy, that is, the allocation of a descriptor to more than one group, has been kept to a minimum. Further, to allow a thematic retrieval of terms thematically related but scattered in different groups, a set of 40 themes have been agreed upon with the EEA and each descriptor has been assigned to as many themes as necessary. Thus, the user can access the thesaurus through the group-hierarchical list, through the thematic list or through the alphabetical list. As a complement to the hierarchical "vertical" relations, an exhaustive series of strong "horizontal" relations between terms have been introduced.

The first published version of GEMET presented 5.298 descriptors, including 109 Top Terms, and 1.264 synonyms in English. This version contained translations funded through different sources into several other EU languages but also others like Russian or the Basque regional language. The GEMET project has been a driver for cooperation with the United States Environment Protection Agency (EPA) since 1999. The development of a "Common Global Environmental Vocabulary" – jointly between EEA, UNEP and EPA was officially announced in February 2000.[3] A "Terminology Project Report"[4] described the links made with the EPA's Terminology Reference System (TRS). Spanish terms in GEMET were of special interest for the American community and thus included into the TRS.

GEMET was shared in different ways over the years. A limited amount of volumes was printed, a PC (Windows) version was developed as well as a web version existed by 2001. As one of the first vocabularies, GEMET content was made available in the format of the Simple Knowledge Organisation System (SKOS)[5] (Miles and Pérez-Agüera, 2007) and since the web version had an API it could be widely accessed and connected to. Later on, queries based on a semantic query language for databases (SPARQL)[6] were added. It was also one of the first sources included into the Linked Open Data (LOD) cloud. After 2001, GEMET was hosted directly by EEA and a dedicated website was developed. In the years following, maintenance focussed on translations. Bulgarian, Czech, Estonian, Russian and Slovenian have been added in 2001.

The addition of Arabic (2009) and Chinese (2010) marked another significant step in internationalisation. Further (regional) languages added in 2012 were Catalan, Slovak, Croatian and Ukrainian. In the same year and for the first time, a smaller set of 15 additional terms was inserted. Thanks to a project with European Neighbourhood countries, Armenian, Georgian and Azerbaijani were included in 2015 and it succeeded in 2017 to include with Icelandic the last European language not yet covered. For selected languages and based on stakeholders' interest, the English definitions were translated over the years. Overall, GEMET terms are today available in 37 languages and after a first substantial term update in 2018 with 100 newly added and translated terms (plus further 200 on a candidate list) it contains close to 5500 terms.

In an Internet survey done in 2000, a strong overall usage of GEMET was identified. Most of the initial thesauri but also other largely thematic thesauri and vocabularies were linked to GEMET through the API. This became more and more popular as the usage of the Internet was growing. It became also clear that usage went beyond data indexing – translation support was of a growing application area. No second study on how this developed over time has been done. The usage of controlled vocabularies to support website translations is an application area, EEA is currently considering.

On the occasion of this article, EEA ran an evaluation of the GoogleAnalytics statistic regarding GEMET usage. Overall, in the year until mid-June 2018, approximately 300,000 user visit were counted. Out of these visits, half are attributed to "organic" search, which includes machine access either through search engines or application programming interfaces (APIs). One quarter each originates to "direct" (a user enters the GEMET URL) and "referral" (another website links to GEMET) usages. Half of the users utilise the English language, which mirrors roughly the geographic distribution of the origination of the users (evaluated through internet protocol (IP) addresses). EEA considers this an excellent uptake of the GEMET service.

Taking a look at the development patterns of GEMET in the past 22 years, there have been technical and political reasons which set priorities and in par ticular triggered the inclusions of languages. The technical demand was steered by the need of EEA to get an overview of the available environmental data in EU Member States. The metadata related to environmental datasets with European relevance was collected and registered in a metadata catalogue. Controlled vocabularies were used to index this metadata. In the countries, some of the the- sauri supported registration functions in their libraries of environmental literature.

Political reasons ranged from so different motivation as the wish to belong to the "European environmental family" and the interest to document national or regional identity in a world with a level of English language dominance to the interest to cooperate stepwise on technical environmental matters before pro- viding eventually later access to environmental data. A main challenge in GEMET development has been governance and content update cost. The governance aspect is related to identifying and running the right processes to identify update needs. Usage by EEA Member Countries has always been diverse – some were rather interested – others did not see a need for a common vocabulary. A good buy-in from the countries is needed to initially agree on an update and later on motivate quality assurance. The latter is one of the reasons why a first content update was only done in 2012, including few key terms, which came into the environmental debate and were missing in GEMET. The strongest example here is "climate change" and its narrow terms.

In 2016, EEA identified the need to start a more substantial content update. Since there is no obvious methodology on how to do that, after some discussion, the following approach was agreed: Circa ten key EEA publications including the 2015 report on the state of environment in Europe as well as the vocabulary used on the EEA website was submitted to the Translation Centre for the European Union (CdT). They did a text analysis resulting into 300 most popular terms and included definitions for them. Out of those, EEA identified 100 terms in 23 languages and submitted them to the contact points in all 28 member states for quality control. After a second round of internal quality control, results were included into a GEMET version 4.1.0. A shortcoming is that the new additions are not available in the remaining 14 non-EU languages.

Considering the effort to systematically involve stakeholder as described above, alternative approaches to collect terms for updating an environmental core vocabulary have been explored shortly after 2000 under the Wikimedia Commons[7] related Wiktionary project. The idea was that users would suggest new terms in a process, which would have needed to be moderated but this did not happen to any useful numbers. Maybe a social media-based approach would be more promising these days on the other hand that could conflict with EEA or member states priorities.

A recurring task has been to define and apply both communalities and the distinction between an environmental core vocabulary like GEMET and domain vocabularies. In the EEA context, these are currently controlled vocabularies for biodiversity, water and climate change adaptation. A relationship can be organised through building thesauri federations. The way to do that technically and methodologically still needs to be identified. One example for orientation to lead out such work is the sematic alignment between EuroVoc – the earlier-mentioned authoritative vocabulary of the European Union – and GEMET published in the EU Open Data Portal[8] through a linked open data (LOD) approach.

A federation solution may be also built using the current implementation of relations in GEMET based on close or full (SKOS) matches to selected national thesauri and to Agrovoc[9] – the well-elaborated thesaurus of the UN Food and Agriculture Organisation (FAO) or to EuroVoc. International thesauri initiatives are welcome to keep using GEMET content through the API or through the vocabulary published in linked open data. Given EEA's resource constraints, it will not be possible to systematically and directly liaise with other communities outside its own Eionet network for content updates or specific technical solutions. Since EEA at the same time is committed to provide reliable web services and has an open data policy, access to GEMET is possible in several user-friendly ways. These measures assure global usage by interested institutions into the future.

2.2 Development and exploration of Chinese-English Environmental Translation Terminology (CEETT)

The UN Sustainable Development Goals provide abstract, overarching governance frameworks (Kanie and Biermann, 2017) which requires important local adaptation and implementation through cross-sectoral collaboration. In the study of the GEMET database, we noticed that the Chinese translations collected in the database developed two decades ago, now only covers a fraction of the large, increasing number of sustainability-related expressions appeared in Chinese official publications and mainstream. With the introduction of translated environmental resources, new locally and culturally-adapted sustainability expressions and terms have been created and used extensively in official and everyday Chinese resources. In order to offer an overall analysis of the growing environmental discourse in China, researchers at the University of Sydney, Beijing Foreign Studies University and Beijing HugeMind Education Technology Ltd developed the

Chinese-English Environmental Translation Terminology (CEETT) which aims at supplementing existing multilingual environmental resources such as GEMET.

The CEETT homepage presents a complete list of all the bilingual sustainability terminologies. Term entries are arranged in an alphabetic order to facilitate browsing and look-up. The system is to a large extent a search engine working on restricted data sources. Users can enter any intended search terms, be it English or Chinese. The system will return a list of bilingual terms containing the user-supplied search string. Fuzzy query is set by default in our system. For instance, when 生产 is used as a search word, such bilingual terms as 本地生产 (Local production), 错峰生产 (Staggered peak production), 能源生产革命 (Energy production revolution), 清洁生产 (Clean production), 全要素生产率 (Total factor productivity), and 生产活动 (Production activities) are returned. The fuzzy query mechanism locates any entries in the database with initial, middle, final part or exact match of the text string 生产. It is the same case with English searches (e.g. "production" as the query word).

Fuzzy search renders CEETT users more relevant terms, which has important applications in environmental translation studies. Terminologies are one of the major hurdles in professional translation and usually not semantically transparent to novice translators. Therefore, explanatory notes (e.g. Domains of Knowledge, Sustainability Development Goals, SDG Implementation Analysis Framework, Sources of Notes, Excerpts of Chinese Texts, and Notes) are available to each entry following the hyperlinks of individual hits. For instance, the notes for 全要素生产率 (total factor productivity) go like 指一个系统的总产出量与全部生产要素真实投入量之比，测算公式为：全要素生产率=产出总量/全部资源投入量 (total factor productivity refers to the ratio of the total output of a system to the actual input of all production factors. The calculation formula is: total factor productivity = total output / total resource input load).

Authorised users can sign in the Admin Centre with credentials to edit the existing terminological expressions. Both the entry proper as well as its explanatory metadata can be modified whenever necessary. New entries can always be added at the Entry Maintenance page of the Admin Centre. The project investigators can add new users and assign relevant roles for them to either contribute new terminological entries or maintain the entire database. Our next planned expansion of the system is to link all the terminologies to parallel concordance lines of environmental corpora or other general-purpose parallel corpora. More real-life usages will inevitably facilitate the mastery the technical terms as well as the idiomatic use in proper translational context.

The basic structure of CEET resembles that of GEMET and the main difference between the two is the direction of translation. While GEMET contains multilingual terminologies translated from English as lingual franca to other languages including Chinese, CEETT encompasses a large and growing number of English translations of environmental expressions that have featured in Chinese official publications and mainstream media, which provide first-hand materials of the current discourse of environmental protection and sustainable development in China. The CEETT was compiled in JavaWeb. The bilingual terminologies were saved as an Excel spreadsheet before they were fed into the web application. The data sheet was then converted to database format to be compatible with the JavaWeb programme. The CEETT is an ongoing project between the University of Sydney and Beijing Foreign Studies University.[10]

This list contains some basic features of the Chinese-English environmental translation database:

1   Translation Mode (how the Chinese sustainability expression was created): literal translation of English terminology (ref. GEMET term base); or locally-created expression; or a mix of both
2   Word Length in Character: 3, 4, 5, or 6, which is an indication of conceptual complexity or cognitive load – long expressions would need to be adapted for public environmental education
3   Domains of Knowledge (comparable to the thematic classification in GEMET)
4   Sustainability Development Goals (SDGs)
5   SDG Implementation Analysis Framework (we created this open-ending list to facilitate policy analysis): Problems, Targets, Standards, Principles, Approaches, Methods, Actions, Actors, Resources
6   Sources of Notes: explanations of the semantic meaning of Chinese sustainability-related expressions
7   Examples: examples taken directly from Chinese official publications
8   Notes: any other notes

The ultimate goal of the database is to assess how local/national sustainability language/culture has been developing after 20 years of the translation and social dissemination of environmental policies (such as the creation of GEMET by the European Environmental Agency). They could be used as matching databases for multilingual databases like GEMET which contain translations from English into other languages.

2.3   Exploring the translation and social dissemination of sustainable terminologies in China

In our study, we used both the GEMET and CEETT databases to study the translation of sustainable terminologies in Chinese official publications and the media, which help us understand the patterns of the growth of sustainable discourse in the country. In our study of original Chinese materials, all data were collated from the Factiva database (Dow Jones 2017), a global news database consisting of a wide range of licensed and free publications in numerous languages including Chinese. The database includes a large range of licensed digital materials published by governmental, industrial and business sources in different countries in their original languages since the mid-1990s. Based on the exploration of the digital Chinese publications, four large groups of terms were extracted which are closely related to sustainable living and citizens' social responsibilities to protect their living and working environments. The four large word categories encompass translations from original English terms and a number of locally-designed expressions representing useful efforts to adapt abstract sustainability principles and goals into concrete and specific actions and behaviours. The four word categories highlighted are responsible behaviours; green ethics and social responsibilities for sustainability; green and sustainable living environments and lastly, green or sustainable transport and travel options. These four sets of sustainability terminologies were used to query the database and extract frequency data of publications on particular topics and dimensions of promoting green living style and the public awareness of sustainability. It is hypothesised that the acceptance and circulation of these translated terminologies by different social agencies may have an impact of the sustainability discourse in China.

The first word category is defined sustainable consumption behaviour: 光盘行动 (guāngpán xíngdòng) (eat-it-up campaign, to avoid food waste); 理性消费 (lǐxìng xiāofèi) (rational consumption); 新文化消费 (xīn wénhuà xiāofèi) (consumption based on new cultures); 低碳消费 (dī tàn xiāofèi) (low-carbon consumption) 低碳消费行为 (dī tàn xiāofèi xíngwéi) (low carbon consumption behavior); 理性购买行为 (lǐxìng gòumǎi xíngwéi) (rational purchase behavior); 过度消费 (guòdù xiāofèi) (over consumption); 环境行为 (huánjìng xíngwéi) (environmental behavior); 环境教育 (huánjìng jiàoyù) (environmental education); 节水 (jié shuǐ) (water saving); 节能 (jiénéng) (energy saving); 节能量 (jié néngliàng) (energy saving); 节能降耗 (jiénéng jiànghào) (energy saving and consumption reduction); 家庭使用 (jiātíng shǐyòng) (family use); 节俭用餐 (jiéjiǎn yòngcān) (thrifty dining); 可持续消费 (kěchíxù xiāofèi) (sustainable consumption); 科学消费 (kēxué xiāo fèi) (scientific consumption); 垃圾分类 (lājī fēnlèi) (waste classification); 铅回收 (qiān huíshōu) (lead recycling); 清洁利用 (qīngjié lìyòng) (clean utilisation); 省电 (shěng diàn) (electricity saving); 消费行为 (xiāofèi xíngwéi) (consumption behaviour); 自家消费 (zìjiā xiāofèi) (home consumption); 自给自足 (zìjǐ zìzú) (self-sufficient); 自我负担 (zìwǒ fùdān) (self-pay); 责任消费 (zérèn xiāofèi) (responsible consumption); 汽车共享 (qìchē gòngxiǎng) (car sharing); and 智慧节能 (zhìhuì jiénéng) (smart energy saving).

The second word category refers to green ethics and social responsibilities for sustainability in Chinese: 碳贞操 (tàn zhēncāo) (carbon chastity); 低碳使命 (dī tàn shǐmìng) (low-carbon mission); 生态文明 (sheng tài wén míng) (ecological civilisation); 生态力量 (shēngtài lìliàng) (ecological power); 社会责任 (shèhuì zérèn) (social responsibility); 社会责任感 (shèhuì zérèngǎn) (social responsibility); 道德选择 (dàodé xuǎnzé) (moral choice); 环境责任 (huánjìng zérèn) (environmental responsibilities); 环保意识 (huánbǎo yìshí) (awareness of environmental protection); 环境伦理 (huánjìng lúnlǐ) (environmental ethics) 环境贡献 (huánjìng gòngxiàn) (environmental contribution); 价值观 (jiàzhíguān) (values); 价值取向 (jiàzhí qǔxiàng) (value orientation); 生产者延伸责任 (shēngchǎn zhě yánshēn zérèn) (extended producer responsibility); 企业社会责任 (qǐyè shèhuì zérèn) (corporate social responsibility); 社会公共利益 (shèhuì gōnggòng lìyì) (social public interests); 诚信经营 (chéngxìn jīngyíng) (business integrity);

消费者态度 (xiāofèi zhě tàidù)(consumer attitude) and 意识改革 (yìshí gǎigé) (awareness reform).

The third word category is green living environment which broadly includes built environments, communities, schools, future city design and sustainable life-style: 百年住宅 (bǎinián zhùzhái) (centennial residence, with energy-efficient and sustainable design); 低碳校园 (dī tàn xiàoyuán) (low carbon campus); 低碳乡村 (dī tàn xiāngcūn) (low-carbon village) 低碳社区 (dī tàn shèqū) (low-carbon community) 低碳家庭 (dī tàn jiātíng) (low-carbon family) 绿色生活 (lüsè shēnghuó) (green life) 零碳生活 (líng tàn shēnghuó) (zero-carbon life) 绿色校园 (lüsè xiàoyuán) (green campus) 绿色设施 (lüsè shèshī) (green facilities); 生态村 (shēngtài cūn)(eco-village); 生态城区 (shēngtài chéngqū)(eco-city); 生态生活 (shēngtài shēnghuó)(eco-life); 生态住宅 (shēngtài zhùzhái) (ecological resi-dence); 生态城市 (shēngtài chéngshì)(eco-cities); 低碳城市 (dī tàn chéngshì) (low-carbon cities) 低碳建筑 (dī tàn jiànzhú) (low-carbon building); 低碳生活 (dī tàn shēnghuó)(low-carbon life); 低碳环境 (dī tàn huánjìng) (low-carbon environment); 海绵城市 (hǎimián chéngshì) (sponge city) 可再生能源建筑 (kě zàishēng néngyuán jiànzhù)(renewable energy building); 绿色建筑(lüsè jiànzhù) (green building); 湿地公园 (shī dì gōngyuán) (wetland park); 消费生活 (xiāofèi shēnghuó) (consumption life); 永续家园 (yǒng xù jiāyuán) (sustainable home); 智慧家庭 (zhìhuì jiātíng) (smart families); 智慧城市(zhìhuì chéngshì) (smart cities); 智能家居 (zhìnéng jiājū) (smart homes); 城市环境 (chéngshì huánjìng) (urban environment); 垂直绿化 (chuízhí lühuà) (vertical greening); 持续环境 (chíxù huánjìng)(sustainable environment); 无废城市 (wú fèi chéngshì) (waste-less city); 智慧社区 (zhìhuì shèqū) (smart communities); 智能城市 (zhìnéng chéngshì)(intelligent cities); and 智能生活 (zhìnéng shēnghuó) (intelligent life).

The last word category studied is green transport and travel options which include high-frequency words such as 共享单车 (gòngxiǎng dānchē) (shared bike); 绿色交通 (lüsè jiāotōng) (green traffic); 零碳交通 (líng tàn jiāotōng) (zero carbon transportation); 绿色快递 (lüsè kuàidì) (green express); 智能物流 (zhìnéng wùliú) (smart logistics); 多式联运 (duō shì liányùn) (multimodal trans-port); 低碳交通 (dī tàn jiāotōng) (low-carbon transportation); 慢行系统 (màn xíng xìtǒng) (slow transportation system); 智能快递 (zhìnéng kuàidì)(smart express); 磁悬浮 (cíxuánfú) (Maglev); 磁悬浮列车 (cíxuánfú lièchē) (Maglev trains); 无碳交通 (wú tàn jiāotōng) (carbon-free traffic); 运输路径 (yùnshū lùjìng) (transport paths); 自动驾驶 (zìdòng jiàshǐ) (autopilot); 智慧交通 (zhìhuì jiāotōng)(smart transport); 低碳出行 (dī tàn chūxíng) (low-carbon travel); 零碳出行 (líng tàn chūxíng) (zero-carbon travel) and 微旅行 (wēi lǚxíng) (mini trips or short, low-cost trips).

2.4 Generalised linear regression models (GLRM) for the social diffusion of sustainability translations

GLRM is widely used in social sciences to explore the relations between a set of explanatory factors and the dependence variable. It is an extension of the ordinary least squares (OLS) regression which assumes the distribution of the dependence variable data to be normal. This study uses GLRM to model the relations between three external factors, that is, sustainability interpreting agents (SIA), sustainability word categories (SWC) and the year of publication (YOP). Sustainability interpreting agents refer to social agents who interpret and adapt locally the abstract principles, idea sets and values of translated sustainability concepts and expressions in the Chinese cultural and social context. For the purposes of illustrating the empirical or formalised analytical models, six social agents which assume the social functions of interpreting, communicating, localising and promoting translated sustainability goals, principles and values were highlighted in the corpus analysis of Chinese digitalised publications on sustainable living and social transition including business sources; official reports; government and political sources; legal sources; major business news sources and top industrial sources.

The sectoral classification framework mirrors the structure of the FACTIVA database. Publications from these sources are intended mainly for specialised audiences with knowledge of and interests in materials that are relevant and significant for their particular sectors, for example, official briefs, governmental and administrative materials, or business and industrial news for professionals. It is hypothesised that the engagement of these sources of information plays an important role in the process of translating, adapting sustainability principles and values in the local context. Using these sources of information in the GLRM can help identify and analyse the relative contributions of these distinct sectors to the discussions around sustainable living and lifestyle change in China over the last 20 years.

The second explanatory variable included in the GLRM construction is sustainability word categories. The last external explanatory variable included in the GLRM is the publication date within the time span of 2000 and 2018. Relevant publication data before 2000 are less than sufficient to build and compare alternative hypotheses using the GLRM. Statistically significant relationships between these sectoral sustainability interpreters are described as multi-sectoral interaction. For example, when using GLRMs to fit and predict publications on sustainability in the large digital database used in this study, if the regression coefficients of the independent variables, that is, the social interpreting agents of sustainability are shown to make contributions in the same direction, multi-sectoral interaction is said to exist among these social actors in their function of the socially-embedded interpretation and communication of sustainability goals and principles within the importing culture and society. Alternatively, if the regression models detect differences between the social actors in terms of the direction of their respective regression coefficients, some lack of multi-sectoral interaction among this group of social interpreters of sustainability, or at least the disengagement of those attributed with negative coefficient scores from the others is ascertained. Higher multi-sectoral interaction provides stronger and better focused social communication network for the effective diffusion of sustainability. Lack of interaction with and engagement of certain sectors can weaken the cross-sectoral consensus and cooperation.

2.5 Exploring impact of translation on the growth of sustainability discourse

This section will explore the impact of the language and knowledge translation processes on the development and wide social diffusion of the sustainable living and lifestyle change discourse in China between 2000 and 2018. The development of formalised linguistic analysis models aligns with previous studies of the cultural and social transmission of ideas and concepts (Cavalli-Sforza and Feldman, 1980). Specifically, in our study, the formalised corpus linguistic analysis will examine whether there is any statistically significant relation between the hypothesised explanatory factors, that is, different word categories of translated sustainability terminology and the six social interpreting agents promoting the social communication of sustainability and the dependent variable which is the growth in the publication of sustainable living related materials and resources in China in the last two decades.

Table 2.1 shows that all of the three explanatory factors, SIA, SWC and YOP have significant impact on the dependent variable which is the publications on sustainable living from various social communication channels. SIA, SWC and YOP are categorical independent variables, and the dependent variable Publications of Sustainable Living is a continuous variable. SIA has six levels ranging from business, governmental to legal sources of information; SWC has four levels referring to four dimensions of the translated terminology of sustainable living which are environmental social behaviour; green ethics and social responsibilities; sustainable community and built environment, and green travel options. Lastly, YOP encloses the two decades between 2000 and 2018. It is necessary to examine variations in the contribution to sustainable living discourse among different levels within each of the three independent variables. This requires the computation of parameter estimates which break the total effect from each independent variable down to each of its component level.

Table 2.2 shows that with the exception of major business news sources, five of the six main sources of information as sustainability interpreting agents have

Table 2.1 Generalised linear regression model (GLRM)_tests of model effects

Source	Type III		
	Wald Chi-Square	df	Sig.
(Intercept)	340.047	1	0.000
Sustainability Interpreting Agents (SIA)	135.819	6	0.000
Sustainability Word Categories (SWC)	160.066	3	0.000
Publication Year (YOP)	225.484	18	0.000

Dependent Variable: Volume of publications containing translated sustainability terminologies
Model: (Intercept), Source of Information, Word Category, Year

Table 2.2 GLRM_ Parameter Estimates_ Sustainability Interpreting Agents (SIA) as IV

Parameter : IV_ Sustainability Interpreting Agents (SIA)	B	Std. Error	95% Wald Confidence Interval		Hypothesis Test		
			Lower	Upper	Wald Chi-Square	df	Sig.
Business sources	468.179	49.5690	371.026	565.333	89.208	1	.000
Official reports	341.455	49.5690	244.302	438.609	47.451	1	.000
Government and political sources	336.521	49.5690	239.368	433.675	46.090	1	.000
Legal sources	244.626	49.5690	147.473	341.780	24.355	1	.000
Major business news	65.271	49.5690	-31.882	162.425	1.734	1	.188
Top industrial sources	-425.031	70.0983	-562.421	-287.641	36.764	1	.000

significant impact on the dependent variable ($P < 0.05$). The largest explanatory factor is Business Sources. The regression coefficient B for this source of information is 468, which suggests that with the increase of one unit in the independent variable, that is, one business source of information, there is an increase of 468 publications on sustainable living in the Chinese materials included in the FACTIVA database. This is followed by Official Reports which has a regression coefficient of 341, indicating that with the increase of one official source of information or report, there is an increase of 341 publications on sustainable living in the same database. To a less extent, Legal Sources and Major Business News Sources also contribute to the growth of the sustainable living discourse over the same time span, that is, from 2000 to 2018. However, the level of contribution from Major Business News Sources is not statistically significant to be included in the GLRM. Top Industrial Sources has been attributed a negative regression coefficient score indicating the disengagement of this social interpreting agency in discussions of transition towards sustainable lifestyle and society building. The lack of alignment of Top Industrial Sources with other sources of information detects a gap in the hypothesised sectoral interaction between social interpreting agents around promoting and building shared social consensus and understanding of sustainability as a priority in socio-economic development.

Table 2.3 shows that with the exception of green transportation and travel alternatives, three of the four categories of the translated sustainability terminology have important contributions to the growing sustainable living discourse in China. The largest contributor or the most significant topic of discussion is the word category of environmentally responsible behaviour with a regression coefficient of 427, suggesting that with the increase of one unit in the independent variable, that is, reporting regarding environmentally responsible behaviour, there is an increase of as many as 427 publications related to this topic in the Chinese publications of the FACTIVA database. Typical examples of translated terms and locally created expressions related to environmentally responsible behaviours include: thrifty dining; sustainable consumption; scientific consumption; waste classification; clean utilisation; save electricity; self-consumption; self-sufficient; self-pay and responsible consumption. The second largest contributing factor is sustainability terms related to green ethics and social responsibilities and duties. The regression coefficient for this level of SWC is 321, which means that with the increase of one unit in this variable, that is, green ethics and social responsibilities, there is an increase of 321 publications across the six sources of information in FACTIVA database. The third dimension of the translated sustainability terminology which contributes significantly to the sustainable living discourse is green living environment such as the design of sustainable built environment. Typical translated and localised terms in this category are millennium housing; low-carbon village; low-carbon community; low-carbon family; green campus; green facility; eco-village; eco-city; eco-life and ecological residence. Some terms in this category refer to abstract concepts such as sustainable community and broader social

Table 2.3 GLRM_ Parameter Estimates_ Sustainability Word Categories (SWC) as IV

Parameter : IV_ Word Category	B	Std. Error	95% Wald Confidence Interval		Hypothesis Test		
			Lower	Upper	Wald Chi-Square	df	Sig.
Environmentally Responsible Behaviour	427.260	37.4669	353.827	500.694	130.044	1	.000
Green Ethics and Social Responsibilities	312.584	37.4669	239.150	386.017	69.605	1	.000
Green Living Environment	107.388	37.4669	33.954	180.822	8.215	1	.004
Green Transportation and travel options	32.429	34.3265	-34.850	99.707	.892	1	.345

environment favourable of sustainable lifestyle. The statistical result suggests that there is not enough interest on green travel options ($p = 0.345 > 0.05$) across the six social interpreting agents.

Table 2.4 shows that the sustainable living discourse saw significant growth only after 2010, which is the first year when the rate of growth of publications discussing the four dimensions of sustainability reached a statistically significant level. Important environmental events in China in 2011 include the publication of the carbon dioxide emission reduction and energy saving legislation as part of the twelfth five-year plan which started in 2011. In this momentous and legally binding regulation, two ambitious emission reduction targets were added which were the country's consumption of non-fossil fuels to reach an overall level of 11.4%; and that a reduction of 17% in carbon dioxide emission per every GDP. This was the first time that the Chinese government officially treated carbon dioxide emission as an integral part of the assessment of the performance of local, provincial and central governmental administrations. Internationally, the Great East Japan earthquake and tsunami trigged the Fukushima nuclear disaster. This caused much concerns and heated debates among the public of food and water safety issues, and a growing public awareness of the importance of sustainable living style, and less dependence on energies that may cause severe environmental problems. Prior to that, only the year 2007 saw important growth of publications on sustainable living, which was the year before the 2008 Beijing Summer Olympics (Beyer, 2008).

2.6 Exploring the interaction among factors of sustainability translation

The GLRM has so far focused on the relations between individual independent variable and the dependent variable. It was found that while all of the three independent variables have significant impact on the growth of the sustainable living discourse in China, contributions from different levels of the three independent variables do vary. For example, within the explanatory variable of Sustainability Interpreting Agents (SIA), Top Industrial Sources have proved to be least engaged in discussions of sustainable living, whereas the other five SIAs have been actively contributing to the interpretation and adaptation of sustainability ethics, principles and idea sets within their sectoral contexts. For example, the SIA which has contributed most to the sustainable living discourse in China is Business Sources. This is followed by Official Reports, Governmental and Political Sources; and to a less extent, Legal Sources.

Within the independent variable of Translated Sustainability Terminology, the word categories of Environmentally Responsible Behaviour and Green Ethics and Social Responsibilities have provided the foci of the discourse of transition to sustainable living in China. It is worth noting that a number of expressions compiled into these two sustainability term categories represent important local adaptation and enrichment of the original English sustainability terminology. Expressions such as eat-it-up campaign (to avoid food waste typically seen in Chinese business

Table 2.4 GLRM_ Parameter Estimates_ Year of Publication (YOP) as IV

Parameter : Year of Publication	B	Std. Error	95% Wald Confidence Interval		Hypothesis Test			
			Lower	Upper	Wald Chi-Square	df	Sig.	
[Year=2018]	470.893	81.6461	310.869	630.916	33.264	1	.000	
[Year=2017]	756.893	81.6461	596.869	916.916	85.940	1	.000	
[Year=2016]	505.929	81.6461	345.905	665.952	38.398	1	.000	
[Year=2015]	457.429	81.6461	297.405	617.452	31.389	1	.000	
[Year=2014]	304.107	81.6461	144.084	464.131	13.873	1	.000	
[Year=2013]	290.242	80.9508	131.582	448.903	12.855	1	.000	
[Year=2012]	292.628	82.4118	131.104	454.153	12.608	1	.000	
[Year=2011]	312.429	81.6461	152.405	472.452	14.643	1	.000	
[Year=2010]	154.750	81.6461	-5.273	314.773	3.592	1	.058	
[Year=2009]	99.643	81.6461	-60.381	259.666	1.489	1	.222	
[Year=2008]	111.714	81.6461	-48.309	271.738	1.872	1	.171	
[Year=2007]	**163.607**	**81.6461**	**3.584**	**323.631**	**4.015**	**1**	**.045**	
[Year=2006]	160.000	81.6461	-.023	320.023	3.840	1	.050	
[Year=2005]	116.786	81.6461	-43.238	276.809	2.046	1	.153	
[Year=2004]	46.750	81.6461	-113.273	206.773	.328	1	.567	
[Year=2003]	28.393	81.6461	-131.631	188.416	.121	1	.728	
[Year=2002]	42.714	81.6461	-117.309	202.738	.274	1	.601	
[Year=2001]	12.429	81.6461	-147.595	172.452	.023	1	.879	
[Year=2000]	16.500	72.0162	-124.649	157.649	.052	1	**.819**	

banquets); thrifty dining; new civilised consumption; self-sufficient; carbon chastity; ecological civilisation; awareness reform are fully embedded in the Chinese culture and social context. Green Living Environment, especially with regard to the development of green communities and sustainable built environment, constitutes another key dimension of the sustainability living discourse in China. Lastly, the corpus data shows that Green Transportation Options represents an under-represented area across the diverse sources of information under study (Kenworthy, 2006).

The yearly distribution data reveal interesting and convincing patterns regarding the wide social promotion of sustainable living in China. Prior to 2011, the growth of publications in this area was rather limited, as there was no significant relation between the independent variable of publication dates and the total volume of publications. The only exception was 2007, the year before the 2008 Beijing Summer Olympics which saw a sudden peak in discussion around sustainable lifestyle and environmental protection. More consistent patterns of the increase in sustainable living and consumption publications were established in the statistical analysis of the corpus data from 2011, another year of major international and domestic environmental events. Internally, the central government published the 12th five-year plan for the period between 2011 and 2015 which incorporated for the first carbon dioxide emission reduction in the overall evaluation of performance of local, regional and central administrations.

This section explores the effects or impact of the interaction between the dual translation processes on the introduction, translation and development of the sustainable living discourse in China. This is based on the hypothesis that the interpretation and adaptation of sustainability principles and values by sectors or social interpreting agents (SIA) may exhibit contrastive or complementary patterns as a result of the sectoral priorities of SIAs for specific dimensions of the sustainability discourse. For example, it is suspected that industrial agents or sources of information may display less interest in topics such as sustainability ethics and responsible consumption behaviour than government and political sources of information. By contrast, major business and industrial interpreting agencies may show stronger interests in promoting sustainable lifestyle such as green travel options in social systems where sustainability innovation is led by industrial or business sectors, instead of governmental agents. Better alignment across sources of information or SIAs in a country in terms of the sectoral interpretation and investment in the sustainable living discourse may serve as an indication of the existence of multi-sectoral interaction which may provide a more favourable and conducive social environment for cross-sectoral cooperation around sustainable development and social transition. By contrast, if the SIAs within a country exhibit distinct patterns of the sectoral engagement with the sustainable living discourse with minimal interaction among the SIAs under study, the disparity thus identified may pose challenges to cross-sectoral partnership around the translation and social adaptation of sustainability principles and values in the national context.

2.7 Interaction of Sustainability Interpreting Agents (SIA) and Sustainability Word Categories (SWC)

The statistical results of the GLRM reported in this section show the effects on the development of the sustainable living discourse of the interaction between different levels of SIAs, that is, the six social interpreting agencies and the four levels of Sustainability Word Categories (SWC), that is, the four dimensions of sustainable living highlighted in this study. The patterns identified in this section provide useful insights into the sectorally-motivated engagement with the sustainable living discourse in China across the six SIAs for the period under investigation that is, between 2000 and 2018, especially over the last ten years when the introduction and adaptation of sustainability principles, values and idea sets evolved gradually from a peripheral position to a key item in the social and economic agenda of the country.

Table 2.5 shows that the Chinese business sector is more engaged in the discussion of environmentally responsible social behaviour and green ethics and values, as there are statistically significant relations identified between the business sources of information or social interpreting agency and the three sustainability terminology categories of behaviour, ethics and sustainable living environment. Secondly, there is no significant relation detected between Chinese business source of information and the word category of transportation and green travel options such as shared bike; green traffic; zero carbon transportation; green express; multimodal transport; low-carbon transportation; slow transportation system; smart express; maglev train (or gaotie in Chinese, high-speed railway); carbon-free traffic and

Table 2.5 Interaction between SIA_SWC: SIA = Business Sources

Parameter	B	Std. Error	95% Wald Confidence Interval		Hypothesis Test		
			Lower	Upper	Wald Chi-Square	df	Sig.
[SIA = Business] * [Word Category = Behaviour]	992.895	112.163	773.058	1212.731	78.362	1	0.000
[SIA = Business] * [Word Category = Ethics]	662.842	112.163	443.006	882.679	34.923	1	.000
[SIA = Business] * [Word Category = Living Environment]	223.842	112.163	4.006	443.679	3.983	1	.046
[SIA = Business] * [Word Category = Transport]	40.105	112.163	-179.731	259.942	.128	1	.721

smart traffic. Similar patterns were found between the sources of information of official reports, government and politics and legal sources of information and the two key dimensions of the Chines sustainable living discourse, that is, environmentally responsible social behaviour and green social ethics and values. The similarities thus identified suggest strong influence from governmental and official sources of information on Chinese business sectors.

2.8   Interaction between Sustainability Word Categories (SWC) and Year of Publication (YOP)

The section explores the distribution of different word categories or discourse dimensions of sustainable living in China between 2000 and 2018. The corpus analyses help reveal the social patterns which underlie the introduction and local adaption of sustainability principles and values over the last two decades.

Table 2.6 shows that publications on environmentally responsible behaviour ranging from domestic waste classification and recycling, collective dinning to household energy consumption began to see significant growth as early as 2006. This suggests that environmentally responsible behaviour represents one of the first and key dimensions of the sustainable living discourse in China (Guarín and Knorringa, 2004). This corpus finding implies that the social interpreting agencies or main sources of information have been deliberately and actively engaging with the public from the start of the green social reform movements. At the governmental level, the fourth five-year plan for public law education and promotion started to engage the public in the development of green communities, green schools and green families as early as 2001. From 2006, environmental regulations, policies and pollution monitoring, management practices provided further incentives to the public participation in the growing environmental debates (Carter and Mol, 2007). These include the publication of the regulations for the public participation in environmental impact assessment and environmental information disclosure and daily air quality monitoring reports for all prefecture-level cities in the country (Lu and Abeysekera, 2014).

Table 2.7 shows that the word category of green ethics and social responsibilities entered into the Chinese sustainable living discourse at a much later stage compared with the first word category of environmentally responsible behaviour. Publications on green ethics and social responsibilities began to see important growth from 2012. This suggests that the social assimilation and establishment of environmental values and principles began to take root the social development discourse in China after half a decade of active government-led education and promotion of environmentally responsible behaviour among the general public. In fact, the social awareness of environmental protection and its impact on public health had reached a record level in 2012 that within the four moth period of July and October, three large collective environmental protects took place in the prosperous southeast coast city Ningbo; Hainan Island in South China Sea and the more remote and socioeconomically disadvantaged city of Shi Fang of Sichuan province, in southwest China. Environmental responsibilities and rights became widely recognised and endorsed by the public (Economy, 2014; Gilbert, 2012).

Table 2.6 Interaction between SWC_YOP: SWC = Behaviour

| Parameter | B | Std. Error | 95% Wald Confidence Interval | | Hypothesis Test | | |
			Lower	Upper	Wald Chi-Square	df	Sig.
[Word Category = Responsible Behaviour] * [Year = 2018]	575.000	164.868	251.864	898.136	12.164	1	.000
[Word Category = Responsible Behaviour] * [Year = 2017]	1059.286	164.868	736.150	1382.422	41.281	1	.000
[Word Category = Responsible Behaviour] * [Year = 2016]	881.571	164.868	558.436	1204.707	28.592	1	.000
[Word Category = Responsible Behaviour] * [Year = 2015]	894.000	164.868	570.864	1217.135	29.404	1	.000
[Word Category = Responsible Behaviour] * [Year = 2014]	640.571	164.868	317.436	963.707	15.096	1	.000
[Word Category = Responsible Behaviour] * [Year = 2013]	575.000	164.868	251.864	898.136	12.164	1	.000
[Word Category = Responsible Behaviour] * [Year = 2012]	669.857	164.868	346.721	992.993	16.508	1	.000
[Word Category = Responsible Behaviour] * [Year = 2011]	709.714	164.868	386.578	1032.850	18.531	1	.000
[Word Category = Responsible Behaviour] * [Year = 2010]	421.286	164.868	98.150	744.422	6.529	1	.011
[Word Category = Responsible Behaviour] * [Year = 2009]	313.571	164.868	-9.564	636.707	3.617	1	.049
[Word Category = Responsible Behaviour] * [Year = 2008]	343.857	164.868	20.721	666.993	4.350	1	.037
[Word Category = Responsible Behaviour] * [Year = 2007]	491.143	164.868	168.007	814.279	8.874	1	.003

(continued)

Table 2.6 (Cont.)

Parameter	B	Std. Error	95% Wald Confidence Interval		Hypothesis Test		
			Lower	Upper	Wald Chi-Square	df	Sig.
[Word Category = Responsible Behaviour] * [Year = 2006]	408.143	164.868	85.007	731.279	6.128	1	.013
[Word Category = Responsible Behaviour] * [Year = 2005]	294.571	164.868	-28.564	617.707	3.192	1	.074
[Word Category = Responsible Behaviour] * [Year = 2004]	135.857	164.868	-187.279	458.993	.679	1	.410
[Word Category = Responsible Behaviour] * [Year = 2003]	93.286	164.868	-229.850	416.422	.320	1	.572
[Word Category = Responsible Behaviour] * [Year = 2002]	110.571	164.868	-212.564	433.707	.450	1	.502
[Word Category = Responsible Behaviour] * [Year = 2001]	52.714	164.868	-270.422	375.850	.102	1	.749
[Word Category = Responsible Behaviour] * [Year = 2000]	29.143	164.868	-293.993	352.279	.031	1	.860

Table 2.7 Interaction between SWC_YOP: SWC = Ethics and Social Responsibilities

Parameter	B	Std. Error	95% Wald Confidence Interval		Hypothesis Test		
			Lower	Upper	Wald Chi-Square	df	Sig.
[Word Category = Ethics] * [Year = 2018]	1073.714	164.8683	750.578	1396.850	42.413	1	.000
[Word Category = Ethics] * [Year = 2017]	1409.714	164.8683	1086.578	1732.850	73.112	1	.000
[Word Category = Ethics] * [Year = 2016]	744.571	164.8683	421.436	1067.707	20.396	1	.000
[Word Category = Ethics] * [Year = 2015]	728.714	164.8683	405.578	1051.850	19.536	1	.000
[Word Category = Ethics] * [Year = 2014]	455.857	164.8683	132.721	778.993	7.645	1	.006
[Word Category = Ethics] * [Year = 2013]	421.714	164.8683	98.578	744.850	6.543	1	.011
[Word Category = Ethics] * [Year = 2012]	340.000	164.8683	16.864	663.136	4.253	1	.039
[Word Category = Ethics] * [Year = 2011]	318.571	164.8683	-4.564	641.707	3.734	1	.053
[Word Category = Ethics] * [Year = 2010]	150.429	164.8683	-172.707	473.564	.833	1	.362
[Word Category = Ethics] * [Year = 2009]	104.286	164.8683	-218.850	427.422	.400	1	.527
[Word Category = Ethics] * [Year = 2008]	121.143	164.8683	-201.993	444.279	.540	1	.462
[Word Category = Ethics] * [Year = 2007]	141.000	164.8683	-182.136	464.136	.731	1	.392
[Word Category = Ethics] * [Year = 2006]	156.714	164.8683	-166.422	479.850	.904	1	.342
[Word Category = Ethics] * [Year = 2005]	128.714	164.8683	-194.422	451.850	.610	1	.435
[Word Category = Ethics] * [Year = 2004]	70.000	164.8683	-253.136	393.136	.180	1	.671
[Word Category = Ethics] * [Year = 2003]	44.000	164.8683	-279.136	367.136	.071	1	.790
[Word Category = Ethics] * [Year = 2002]	53.286	164.8683	-269.850	376.422	.104	1	.747
[Word Category = Ethics] * [Year = 2001]	36.000	164.8683	-287.136	359.136	.048	1	.827
[Word Category = Ethics] * [Year = 2000]	21.857	164.8683	-301.279	344.993	.018	1	.895

Table 2.8 shows that compared with the previous two lexical categories, the sub-class of Chinese translated sustainability terminology which promotes the public participation in and contribution to the construction of sustainable, green living environments, both physically and conceptually only began to grow at a significant level from 2016 onwards. This dimension of the sustainable living discourse thus represents a new and emerging area which is very likely to see important growth in the coming years, as the Chinese government and business sectors invest more in the construction of sustainable housing and social facilitates such as community green space, ecological residence, renewable energy building, smart homes, vertical greening, as well as the development of green urban development policies and strategies.

2.9 Conclusion and future research

This chapter provided corpus-based discourse analyses of the main social factors which contribute to the introduction and cultural adaptation of values, principles and idea sets of environmental sustainability in China. Using a combination of qualitative and quantitative methods to explore large original Chinese publication databases, this study illustrated the changing patterns and mechanisms underlying the social dissemination and adaptation of sustainability materials in China, providing first-hand evidence of the country's sustainable development strategies and agendas. In terms of methodological innovation, this study integrated qualitative and quantitative research methodologies from translation studies, Chinese media and discourse analysis and quantitative social sciences. It offered valuable insights into the different stages of the production and development of sustainability translation resources, and the social dissemination and the subsequent utilisation of translated sustainable values, concepts and principles in China over the last two decades. The empirical corpus analytical models constructed illustrate the social mechanisms that underlined the social diffusion of translated sustainability materials; and the intra-sectoral framing of the industrial materialisation of sustainability goals and principles. These innovative models were introduced to Chinese environmental translation studies; and can be effectively adapted for the study of the translation and social communication of sustainability in other social and cultural systems, as we witness global trends of transitioning towards sustainable societies and communities. As with many empirical studies, while this study has identified and developed approaches to address key research questions related to the social diffusion of sustainability translation resources in China, it has also raised new questions that remain to be answered in future research, for example, whether the features, patterns and mechanisms of the social translation, dissemination and industrial materialisation of sustainability aims, goals and principles are unique and exclusive to the Chinese society and cultural system. In other words, mechanisms such as whether the sectorally-motivated framing of sustainability principles which have been found in the study of Chinese environmental discourse may also be observed in other countries and/or regions. If this hypothesis hols true, it will provide a strong theoretical basis for the development of analytical methods and procedures to compare different countries and

Table 2.8 Interaction between SWC_YOP: SWC = Sustainable Living Environment

Parameter	B	Std. Error	95% Wald Confidence Interval		Hypothesis Test		
			Lower	Upper	Wald Chi-Square	df	Sig.
[Word Category = Sustainable Living Environment] * [2018]	231.571	164.8683	-91.564	554.707	1.973	1	.160
[Word Category = Sustainable Living Environment] * [2017]	468.143	164.8683	145.007	791.279	8.063	1	.005
[Word Category = Sustainable Living Environment] * [2016]	353.000	164.8683	29.864	676.136	4.584	1	.032
[Word Category = Sustainable Living Environment] * [2015]	209.429	164.8683	-113.707	532.564	1.614	1	.204
[Word Category = Sustainable Living Environment] * [2014]	143.000	164.8683	-180.136	466.136	.752	1	.386
[Word Category = Sustainable Living Environment] * [2013]	170.143	164.8683	-152.993	493.279	1.065	1	.302
[Word Category = Sustainable Living Environment] * [2012]	200.286	164.8683	-122.850	523.422	1.476	1	.224
[Word Category = Sustainable Living Environment] * [2011]	224.429	164.8683	-98.707	547.564	1.853	1	.173
[Word Category = Sustainable Living Environment] * [2010]	94.429	164.8683	-228.707	417.564	.328	1	.567
[Word Category = Sustainable Living Environment] * [2009]	35.000	164.8683	-288.136	358.136	.045	1	.832
[Word Category = Sustainable Living Environment] * [2008]	37.286	164.8683	-285.850	360.422	.051	1	.821
[Word Category = Sustainable Living Environment] * [2007]	71.571	164.8683	-251.564	394.707	.188	1	.664
[Word Category = Sustainable Living Environment] * [2006]	122.857	164.8683	-200.279	445.993	.555	1	.456
[Word Category = Sustainable Living Environment] * [2005]	100.857	164.8683	-222.279	423.993	.374	1	.541
[Word Category = Sustainable Living Environment] * [2004]	38.143	164.8683	-284.993	361.279	.054	1	.817
[Word Category = Sustainable Living Environment] * [2003]	35.143	164.8683	-287.993	358.279	.045	1	.831
[Word Category = Sustainable Living Environment] * [2002]	60.143	164.8683	-262.993	383.279	.133	1	.715
[Word Category = Sustainable Living Environment] * [2001]	18.571	164.8683	-304.564	341.707	.013	1	.910
[Word Category = Sustainable Living Environment] * [2000]	7.571	164.8683	-315.564	330.707	.002	1	.963

communities regarding their transition towards sustainable societies with distinct local features and characteristics, as a result of the locally-based and culturally-rooted interpretation and translation of sustainability.

Notes

1 www.eionet.europa.eu/gemet
2 http://eurovoc.europa.eu/
3 EPA press release https://archive.epa.gov/epapages/newsroom_archive/newsreleases/0058lb9c5f18ca738525687a006dcb6d.html
4 Terminology Project Report, Joel Tochterman and Valorie Lee, EPA unpublished report May 2000
5 www.w3.org/2004/02/skos/
6 www.w3.org/TR/sparql11-query/
7 https://commons.wikimedia.org/wiki/Main_Page
8 https://data.europa.eu/euodp/en/data/dataset/eurovoc_gemet
9 http://aims.fao.org/
10 https://translation-terminology.sydney.edu.au

References

Beyer, S. (2006) The Green Olympic Movement: Beijing 2008, *Chinese Journal of International Law* 5(2): 423–40, https://doi.org/10.1093/chinesejil/jml018.

Budin, G. (2007) Semantic Systems Supporting Cross-Disciplinary Environmental Communication. In: O. Hryniewicz, J. Studzinski, and A. Szediw (eds.) *Environmental Informatics and Systems Research*, vol. 2: Workshop and Application Papers. Warsaw EnviroInfo Conference 2007. Aachen, Shaker Verlag, pp.23–7.

Carter, N. and A. P. J. Mol (eds.) (2007) *Environmental Governance in China*, London: Routledge.

Cavalli-Sforza, L. L. and M. W. Feldman (1980) *Cultural Transmission and Evolution: A Quantitative Approach*. Princeton, NJ: Princeton University Press.

Economy, E. (2014) Environmental governance in China: State control to crisis management. *Daedalus* 143:2, 184–97.

Gilbert, N. (2012) Green protests on the rise in China: Environmental groups use momentum to push for reforms. *Nature*, 488: 261–2.

Guarín, A. and P. Knorringa (2014) New middle-class consumers in rising powers: Responsible consumption and private standards. *Oxford Development Studies* 42:2, 151–71.

Jones, D. (2017) Factiva Database. Dow Jones Incorporated.

Kanie, N. and F. Biermann (eds.) (2017) *Governing through Goals: Sustainable Development Goals as Governance Innovation*, Cambridge, MA: MIT Press.

Kenworthy, J. R. (2006). The eco-city: Ten key transport and planning dimensions for sustainable city development. *Environment and Urbanization* 18(1), 67–85.

Lu, Y. and Abeysekera, I. (2014). Stakeholders' power, corporate characteristics, and social and environmental disclosure: evidence from China. *Journal of Cleaner Production*, 64, 426–36.

Miles, A. and J. R. Pérez-Agüera (2007) SKOS: Simple Knowledge Organisation for the Web. *Cataloging and Classification Quarterly*, 43:3–4, 69–83.

Steinberger, R., B. Pouliquen and J. Hagman (2002) Cross-lingual document similarity calculation using the multilingual thesaurus EUROVOC. In: A. Gelbukh (eds), *Computational Linguistics and Intelligent Text Processing*. CICLing 2002. Lecture Notes in Computer Science, vol. 2276. Springer, Berlin, Heidelberg.

3 The Environmental Thesauri of CNR EKOLab

Sabina Di Franco, Diego Ferreyra, and Paolo Plini

3.1 Introduction

This chapter aims to illustrate the activities of the Environmental Knowledge Organisation Laboratory (EKOLab) of CNR-IIA and the present status of the Environmental Thesauri developed by our group. The Environmental Knowledge Organisation Laboratory (EKOLab) of CNR-IIA started its activities in the field of environmental knowledge organisation in 1992. Activities are related to the construction and maintenance of mono- and multi-lingual thesauri, classification schemes and terminological systems for science and the environment. At the present time, EKOLab has four fully developed thesauri available online: EARTh (Environmental Applications Reference Thesaurus) on environmental terminology, EOSterm (3200 terms on Earth Observation and Remote Sensing, Snowterm (3500 terms on the cryosphere), GeothermThes (about 2000 terms on geothermal energy). Two other thesauri are quite complete but not online at the moment, the first one on natural hazards that contains more than 2000 concepts and PollThes with about 7000 terms on air pollution. The thesauri are at least bilingual (English and Italian) and all of them contain relations between terms. Three of them – Snowterm, EOSterm and GeothermThes – also include glossaries. Moreover, over the years we have developed classification schemes, ontologies, monolingual vocabularies and other terminological "tools" for specific projects, including EARTh, which contains about 14,500 terms, more than 13,000 relationships and about 5000 definitions, is a general-purpose thesaurus for the environment and was developed since 2001 by CNR aiming at creating a new thesaurus for the environment. It extends the GEMET – the *de facto* standard when speaking of a general-purpose thesaurus for the environment in Europe – content and revises its categorical and thematic structure.

3.2 The need for a sound environmental terminology

It is increasingly evident that, in spite of relevant political and technological changes over the past few decades, no apparent impact has been made to reverse the continuing decay of the global environment. Therefore, the condition of the state of the environment, its problems and the solutions that humankind can implement for sustainable interaction, are important and urgent issues that must be addressed. The impact that correct information to scientists, decision makers, NGOs and the general public can have for coping with the environmental problems, is also quite clear. The sound and relevant information required by all these players are difficult to locate and deliver unless suitable means of control of the information are not developed, disseminated and used. Therefore, information control systems based on knowledge organisation and intelligent analysis are of paramount importance.

The internet "information deluge" is a continuous and rather chaotic flow, hence how to give an answer to the increasing need for clear, and trustworthy information on the environment? How to share knowledge? How to support a citizen-science perspective? Which are the best practices to switch towards a new resilient information ecosystem? The challenge is to find models and tools to build an open and structured knowledge to facilitate access to validated and reliable information. Language plays a fundamental role in this framework. The confusion and misunderstanding on scientific terms are often underestimated in their consequences. It is needed to start back from "words," from their meaning and relations between concepts and terms, as precise comprehension allows a more "precise" behaviour. Furthermore the language is in itself rich of semantic ambiguity and polysemy and the meaning of each term has a high degree of context-dependency, for example, the word "mercury," depending on context refers to a planet, a chemical element, a Roman god or a famous rockstar. These statements are true in all communication areas and science communication and information are no exception to these rules. Actually, in science, we need to share information and data with colleagues, to store and search datasets, to translate in other languages, to teach, to inform and persuade public, stakeholders, media and policymakers. We must also consider that in the last few years the complexity and plurality of the disciplines related to sciences, and environment in general, have increased. The domains of disciplines that deal with environmental sciences, form a complex interlinking network and are frequently overlapping: mutual understanding is sometimes taken for granted whereas each discipline has its specific jargon. For example, a geotechnical engineer will use the term "soil" having in mind properties such as density, porosity and resistance while an agronomist will use this term considering the organic content and fertility.

Whilst the need for a correct and precise vocabulary is an easily understandable issue the efforts to improve the state of play are more time and resources consuming. Terminological tools, be they controlled vocabularies, thesauri, glossaries, ontologies could help in these tasks. Thesauri, in particular, are widely employed as common ground among the different communities working in environment-related domains: they allow users to share and agree upon scientific/technical terms in the target domain and to express them in multiple languages. Controlled vocabularies and thesauri have been deployed by different communities having a large spectrum of competencies. They have been created embodying specific points of view and based on different ways of conceptualisation.

The experience gained by the CNR, from its work on a general multilingual environmental thesaurus and accessory terminologies, has shown that, in order to provide an answer to the need for sound and relevant information, it is necessary to take advantage not only of the terminology and thesaurus disciplines but also of elements of linguistics, artificial intelligence, knowledge organisation, logic (why not Aristotelian?) and a multitude of scientific disciplines, including information science. In other words, severe problems require the mobilisation of a variety of intellectual resources, a transversal, trans-disciplinary approach, and the implementation of open, intelligent strategies.

3.3 The history of EKOLab

In the late 1980s, CNR had the idea of a reference language for the environment – a language that could be shared at least at the European level. The terminological resources available at that time included the Multilingual Descriptor System (MDS) of the European Commission comprised of 1400 terms in six European languages, edited in 1983, and a short UNEP Infoterra list of less than 200 terms, edited in 1984. In 1990, the monumental contribution by Paenson appeared – *Environment in Key Words: A Multilingual Handbook of the Environment,* comprised of 4100 terms in four languages and the first Infoterra thesaurus with 1200 descriptors in five of the languages of the UN systems. It is interesting and regretful that both MDS and in particular Paenson's work – the fruit of 18 years of dedicated effort – were not widely marketed or disseminated among the user community and therefore received little acknowledgement. Only the MDS was further developed by CNR in 1989 in a Bilingual Descriptor System, by merging the English and Italian terms of MDS with 144 terms of the Infoterra thesaurus 1984 and with 500 CNR terms. The Bilingual Descriptor System, like MDS, had no thesaurus structure but its added value was an improved, sophisticated thematic and sub-thematic classification scheme. In the early 1990s, while monolingual environmental thesauri were being developed in Germany, the Netherlands, Spain and France, the only reference multilingual thesaurus was the Infoterra thesaurus. Since 1989, CNR had started work on the development of a parallel environmental thesaurus, based on the Dutch Milieu-thesaurus. This thesaurus was chosen since its structure and size (2369 terms) were deemed satisfactory. The 2369 terms were very well structured in a thesaurus system made up of a core thesaurus and a series of thesaurus-structured lists of national interest. The Milieu-thesaurus was translated into English, Italian and German were published as CD-ROM by CNR in 1993.

In 1994, in the context of collaboration with UNEP's Infoterra working group, CNR translated into Italian the Infoterra thesaurus and disentangled its structure as a preparatory step for the participation of CNR in the compilation of the UNEP-Infoterra EnVoc, edited in 1997. In 1995, in collaboration with the Dutch institution responsible for the Milieu-thesaurus, CNR produced on behalf of the Task Force of the European Environment Agency (Brussels), the MET, Multilingual Environmental Thesaurus and a Classification Scheme for the same thesaurus. The MET contained the 2300 terms of the Milieu-thesaurus translated into eight languages.

In 1994, in the context of collaboration with UNEP-Infoterra, CNR translated into Italian the Infoterra thesaurus and disentangled its structure as a preparatory step for the participation of CNR in the compilation of the UNEP-Infoterra EnVoc, edited in 1997. In 1995, in collaboration with the Dutch institution responsible for the Milieu-thesaurus, CNR produced on behalf of the Task Force of the European Environment Agency (Brussels), the MET, Multilingual Environmental Thesaurus and a Classification Scheme for the same thesaurus. The MET contained the 2300 terms of the Milieu-thesaurus translated into eight languages.

In 1996, CNR and other thesaurus custodians convinced the EEA, which in the meantime was moved from Brussels to Copenhagen and did not intend to adopt (even temporarily) the MET, to launch a new, broader initiative for the development of a multilingual thesaurus. This thesaurus had to include 2000 terms selected from the Umwelt Thesaurus, 2300 terms of the Italian CNR Thesaurus, 2000 terms of the MET Classification Scheme, the whole MET, the UNEP-Infoterra EnVoc, as well as selected portions of the Spanish Tesauro de Medio Ambiente and the French Lexique Environnement Planète.

The idea was accepted to merge in a unique thesaurus the best terminologies of the six existing multilingual thesauri, in order to propose a reference thesaurus to the different (European) organisations and to provide an agreed common language basis for the exchange of environmental information. The task of compiling the thesaurus was delegated to CNR, Rome and UBA (Umweltbundesamt), Berlin. In order to ease the translation burden and to obtain an exact correspondence between the concepts expressed in different languages and the English representation of the terms, it was decided to complement almost all the terms of the thesaurus with a definition in English, obtained by an authoritative source.

The thesaurus was prepared to satisfy the three functions:

- System of controlled terms
- Multilingual dictionary
- Glossary

In 1998, a preliminary version of the thesaurus was tested and interactively applied in a successful way to an Italian environmental information system of national extension. In 1998, the whole terminology was made available in British English, American English, Italian and German, and translation to the other European languages was performed by the EEA and by the EEA's Topic Centre for the Catalogue of Data Sources ETC/CDS (which was coordinating the application of the thesaurus).

In 1999 the final version of GEMET was published as a tool of the EEA's CDS: a thesaurus of 5300 preferred terms in 12 languages and 1200 synonyms in English. At present, GEMET is applied to the Catalogue of Data Sources of the EEA's ETC/CDS and as a multilingual dictionary and it is used in the search engine of the EEA website.

GEMET was conceived as a "general" thesaurus, aimed to define a common language, a core of terminology for the environment. GEMET was compiled by merging the terms of several multilingual documents. The merging was performed both on conceptual and formal basis. Coinciding concepts in the different thesauri were identified and scored. From the point of view of translations, GEMET provides a complete equivalence (all the descriptors have an equivalent in a different language) and is a *de facto* standard of environmental thesauri. Once concluded the work on GEMET, the EKOLab work in the field of knowledge organisation related to terminology and thesauri continued.

During the 1999 CNR, EEA, UNEP and US EPA were engaged in a dialogue to establish a project to develop a global environmental thesaurus. UNEP's ENVOC thesaurus had broad multilingual coverage but narrow content coverage. EEA's GEMET thesaurus had broad content coverage but narrow multilingual coverage (European languages only). The idea of a comprehensive global thesaurus was proposed by US EPA who foresaw the possibility and opportunity to translate an improved thesaurus based on the multilingual thesaurus GEMET of the EEA, that CNR had in the meantime proposed to US EPA, into languages of the APEC area. From the outset, CNR, owing to the past collaboration with UNEP-Infoterra on the EnVoc thesaurus, as well as with US EPA and EEA, favoured this idea of a new global thesaurus. The four lead organisations met in Santa Fe, New Mexico in January 2000 and through a quadrilateral agreement, called the Santa Fe Agreement decided to pool experience and expertise and mobilise external partners to develop a multi-thematic, multilingual vocabulary of the environment.

The CNR played an important role in the development of the UNEP's ENVOC thesaurus (formerly known as the Infoterra thesaurus). The modular components of T-REKS©, Thesaurus-based Reference Environmental Knowledge System© release 2000, developed by CNR were used to provide the UNEP-Infoterra network with a core thesaurus module that was compiled by CNR with the assistance of UNEP.

In 2000 starting from T-REKS© EKOLab started the development of a thesaurus for the environment called EARTh (Environmental Applications Reference Thesaurus). The thesaurus is constantly updated following the growing request of an up-to-date terminology in relevant environmental topics, it contains about 14,500 terms, in English and Italian. It is provided with hierarchical relations and classified according to a matrix system. At the present EKOLab develops and maintains other thesauri EOSterm (Earth Observation and Remote Sensing), Snowterm (snow and cryosphere), GeothermThes (geothermal energy).

EOSterm was developed starting from a project aiming at developing a controlled and structured terminology system related to earth observation, Geographic Information System and remote sensing. The first phase corresponded to the identification of potential sources of terminology, the selection of proper terms, their extraction and the creation of a terminology database, terms have been classified according to the EARTh structure. The semantic classification foresees the allocation of each term in a relation tree starting from the more general concept represented by the category. The result is a English-Italian terminology system containing about 3200 terms. The EOSterm has been updated with recent satellites and sensors platform terms.

The Snowterm thesaurus is an example of a structured reference multilingual scientific and technical vocabulary, covering the terminology of a specific knowledge domain in the polar and the mountain environment. The thematic areas, covered at present, deal with snow and ice physics, snow and ice morphology, snow and ice radiometry, remote sensing and GIS applied to cryosphere environment, sea ice, avalanches, and glaciers. At present, the database contains around 3500 terms.

GeothermThes was developed in a project with CNR-IGG (Institute of Geosciences and Earth Resources) as part of an information system on geothermal resources. It contains 2000 concepts in English and Italian.

Two other thesauri are complete but not online at the moment, the first one on natural hazards and emergency management that contains more than 2000 concepts and PollThes with about 7000 terms on air pollution, pollutants and the related health issues.

The thesauri are published and managed online with Tematres an open source vocabulary server, a web application to manage and exploit vocabularies, thesauri, taxonomies and formal representations of knowledge, developed by Diego Ferreyra from Centro Argentino de Información Científica y Tecnológica, CAICYT-CONICET. This open source vocabs management system allow to manage relation between concepts, define and create relations between concepts and any Internet entity, use web services to synchronise and notify changes, use JSON and Linked Data: JSON-LD, share changes and terminological news via RSS, enables the export of controlled vocabularies in many metadata schemas: Skos-Core (Simple Knowledge Organization System), BS 8723 (Structured Vocabularies for Information Retrieval), Dublin Core (ISO 15836-2003), MADS (Metadata Authority Description Schema), TopicMaps (ISO/IEC 13250:2003), IMS VDEX Scheme (Vocabulary Definition and Exchange) WXP WordPress XML, TXT, SQL, Zthes.

Our approach is based on the need to share a common and stable meaning of specific terminology to help communication both within a community and to link different group of the society through clear and unambiguous language. The goal is the development of terminological tools that aim to become advanced tools to be applied in environmental information management.

3.4 The thesauri

Thesauri are widely employed as common ground enabling communication among the different communities working in environment-related domains: they allow users to share and agree upon scientific/technical terms in the target domain and to express them in multiple languages. In recent years, several structured controlled vocabularies (thesauri) have been deployed by different communities having a large spectrum of competencies. They have been created embodying different points of view based on different ways of conceptualisation and have their specific ISO standards. Their development reflects different scopes and implies quite a range of levels of abstraction and detail.

The thesaurus plays an important role as a tool for the conceptual and formal control of the information flow concerning the environment. It will also be useful as the link element between different environmental information systems and the catalogues of similar national structures. From the thesaurus, it will be possible to access the databases of any institution responsible for data that has adopted the thesaurus for the management of its environmental information system. Thesaurus could also be a useful instrument for the compilation of the reports on the state of the environment and for the organisation of libraries and documentation centres.

In general, it can be assumed that thesauri are a reference tool for librarians, documentalists, database developers, thesaurus developers, terminologists, geographic information system (GIS) specialists, translators, interpreters and the broad spectrum of environmental information users.

The thesaurus core can be linked to complementary extensions containing terms that are peculiar to a country or region – mainly organisational attributes (ASEAN, SADC, OAS, GCC, etc.) and geographic attributes (Zambezi basin, Amazon Delta, Great Barrier Reef, etc.) but also some environmental terms (Highveld ecosystems, Sorghum, Customary Law, etc.) and terminologies needed by the environmental impact statements (EISs): standard lists of international usage; standard lists of national or local use; specific international, national, local terminologies; and so on.

It is obvious that the dimension of the global thesaurus can be limited only by three criteria and operations:

1   The exclusion of terms not pertinent because too general
2   The scattering in accessory lists of terms not pertinent because too specific or of too local interest
3   The linking of these specific lists/extensions to already existing, if and where existing, authoritative lists

3.5 EKOLab semantic model

EARTh (Environmental Application Reference Thesaurus) and the other thesauri are based on a multidimensional classificatory and semantic model. The "vertical structure" of the Thesaurus is the fundamental constituent of such a model. This structure is basically mono-hierarchical. It has been developed according to a tree semantic model and is founded on a system of categories. It is organised in a framework composed of different levels and classification knots and comprises hierarchical relationships.

The notion "category" has taken on different meanings in the history of Western thinking; from the ontological and logical point of view to linguistics, from philosophy to semiotics to psychology. In the classification science categories have also been considered as the foundation (not always visible) of knowledge organisation systems and are utilised for different purposes. Categories are here conceived in their primitive Aristotelian form as the most general genera or the logical progenitors under which every single term can be placed.

The first two levels of the classification correspond to the system of categories. The first level includes four "super-categories": Entities, Attributes, Dynamic aspects and Dimensions. Entities constitute "things." Attributes define the character of "things," at least in their static aspects. Dynamic aspects related to transformations and operations connected to "things." Dimensions identify the spatiotemporal circumstances where all this is manifested.

In the subsequent level of the classification, the super-category Entities is divided into Material entities and Immaterial entities. Attributes include three different categories: Properties; Structure and Morphology; Composition. Dynamic aspects comprehend: Processes; Conditions; Activities. Dimensions refer to Space and Time.

According to this model, the semantics of the terms is, in fact, described by the categories where they are located. Following a bottom-up perspective, terms could be analysed according to a progressive hierarchical scale. In that scale, conceptual features are progressively discarded following an intensional perspective, while in an extensional perspective the number of things associated to that intension is increased. The maximum level of generality is thus reached. Categories represent the top of this vertical structure that analyses the meaning of the terms according to a logical perspective.

It can be considered as an operative tool that – by providing an interpretation of the meaning of the terms and by placing them in the classificatory-hierarchical tree – aims to orientate the users towards the most "essential" characteristics of their semantics. Nevertheless, it does not limit the conceptual analysis of terms to a static and univocal view. Awareness of the semantic complexity associated with each term is maintained. Different layers of meaning have to be explored, even if there is a hierarchy of semantic traits and each one of them contributes to lexical signification with a different specific weight.

3.6 Conclusion

At the beginning of our work we thought and, in a certain way still do, that environment was mainly analysed with a static and sectoral approach, reflecting a vision pertaining to classic science and to environmental policy with limited openings to the development of renewed approaches and methods to analyse the environmental complexity.

Moreover, there is a constant need to follow the terminological changes in science and technology (new terms and/or new meanings, new topics and issues). Starting from this premise, we adopted a more inclusive approach concerning both conceptual coverage and semantic organisation, and taking also into considerations suggestions arising from the development of applied ontologies, we started to work on an environmental thesaurus format that contains some innovative elements. We strongly believe in the importance of sharing knowledge and in a clear and precise use of language. Terminology and its tools can help the mutual understanding especially in a complex and multidisciplinary field such as environmental sciences.

The development of a common and shared environmental terminology is still influenced by at least four factors: a global vision to break down linguistic boundaries to access to sound and relevant information on environmental matters; the availability of qualified and motivated personnel; he availability of suitable financial resources, which in turn depends on the interplay between demand and offer of a quality product; and, last but not least, the commitment of environmental agencies and ministries of the environment to provide national level input to support a global initiative.

References

Aitchison, J., Gilchrist, A., and Bawden, D., 2003. *Thesaurus Construction and Use: A Practical Manual*. Fourth edition. Published by Aslib IMI, The Association for Information Management, Staple Hall, Stone House Court, London. This edition published in the, Taylor & Francis e-Library, 2005.

Albertoni R., De Martino, M., Di Franco, S., De Santis, V., and Plini, P., 2014. EARTh: An Environmental Application Reference Thesaurus in the Linked Open Data Cloud. *Semantic Web*, 5, pp. 165–171. DOI 10.3233/SW-130122.

de Lavieter, L. (ed.), 1995. Multilingual Environmental Thesaurus. Part 1, English; Part 2, Français; Part 3, Deutsch; Part 4, Nederlands; Part 5, Italiano; Part 6, Norsk; Part 7, Dansk; Part 8, Español. NBOI, Nederlandse Bureau voor Onderzoek Informatie / EEA-TF – European Environment Agency – Task Force, Amsterdam, November 1995, pp. (English) vi, A-78; B-112; C-56; D-199, total 445.

Di Franco, S., Rapisardi, E., and Giardino, M., 2015. From "information deluge" to explicit knowledge: How web technologies and web collaboration could support Natural Hazards Communication. *Geophysical Research Abstracts*, Vol. 17, EGU 2015–4783, EGU General Assembly 2015.

EEA, European Environment Agency, 1999. GEMET, GEneral Multilingual Environmental Thesaurus. pp. ix, Volume 1: Systematic List of Descriptors, pp. 44; Volume 2: Thematic List of Descriptors, pp. 78; Volume 3: Alphabetical List of Terms, pp. 550; Volume 4: Concordance List, pp. 127; Volume 5: Multilingual List of Descriptors, pp. 536. EEA, Copenhagen, August.

Enderman, J. C., de Lavieter, L., and van Vloten, A. A., 1990. Milieu-thesaurus. Systeem van gecontroleerde termen voor het ontsluiten van milieu-informatie. pp. iii, 116, App. s'Gravenhage. ISBN: 90-346-2161-8.

Felluga, B. (ed.), 1995. *Multilingual Thesaurus for the Environment. Classification Scheme*. Roma: CNR, pp. 3, 90.

Felluga, B., Pàlmera, M., and Lucke, S., 1989. *Sistema Bilingue di Descrittori per l'indicizzazione, la categorizzazione e la codificazione dei termini ambientali. Bilingual Descriptor System for Indexing, Categorizing and Codifying Environmental Terms*. Roma: CNR-IIBM, pp. xxiii, 278.

Felluga, B., Lucke, S., Pàlmera, M., Plini, P., de Lavieter, L., and Deschamps, J. (eds.), 1994. Thesaurus per l'ambiente – Versione quadrilingue/Thesaurus for the Environment – Quadrilingual Version/Milieuthesaurus – Viertalige vertaling/Thesaurus für die Umwelt, CNR-SIAM & CNR-UPIS edition on CD-ROM, Milan, 1994.

Ferreyra, D., and Bosch, M., 2013. *Vocabularios controlados para la comunicación científica.* Communication in Conference. Encuentro Nacional de Catalogadores, 4. Biblioteca Nacional Argentina Buenos Aires.

International Organization for Standardization, 2011. ISO 25964-1:2011, information and documentation. Thesauri and interoperability with other vocabularies. Part 1: thesauri for information retrieval. Part 2: interoperability with other vocabularies. Geneva. International Organization for Standardization.

Mazzocchi, F., De Santis, B., Tiberi, M., and Plini, P., 2007. Relational Semantics in thesauri: Some remarks at theoretical and practical levels. *Knowledge Organization*, 34 (4), pp. 197–214.

Ministère de l'environnement, 1995. Lexique environnement – Planète. Tome 1, Liste alphabétique, pp. 83; Tome 2, Liste thématique, pp. iv, 186. Ministère de l'environnement, Paris, December, 1995.

MOPTMA, Ministerio de Obras Públicas, Transportes y Medio Ambiente, 1990. Tesauro Multilingüe de Medio Ambiente. MOPTMA, Madrid, 1995, CD-ROM Edition, 1995. (Includes the contents of: MOPU, Ministerio de Obras Publicas y Urbanismo. Tesauro de medio ambiente. MOPU, Madrid, 1990, pp. xxxii, 319.)

Nativi, S., Mazzetti, P., and Craglia, M., 2017. A view-based model of data-cube to support big earth data systems interoperability. *Big Earth Data*, 1, pp. 75–99.

Paenson, I., 1990. *Environment in Key Words. A Multilingual Handbook of the Environment.* Oxford: Pergamon Press. Vol. 1, pp. xxxiv, 662. Vol. 2, pp. 268. ISBN: 0 08 024524 2

Plini, P., Di Franco, S., and De Santis, V., 2009. A state-of-the-art of Italian National Research Council (CNR) activities in the area of terminology and thesauri. Proceedings of the European conference of the Czech Presidency of the Council of the EU TOWARDS eENVIRONMENT.

Plini, P., and Felluga, B., 2000. T-ReKS, Thesaurus-based Reference Environmental Knowledge System. TDNet, Conférence pour une infrastructure terminologique en Europe. 13–15 March 2000, Maison de l'UNESCO, Paris, France.

Plini, P., Felluga, B., Lücke, S., and Pàlmera, M., 1994. INFOTERRA Thesaurus di termini per l'ambiente. Thesaurus of Environmental Terms. Versione quadrilingue italiano-inglese con equivalenti linguistici in francese e spagnolo della terza edizione. Ministero dell'Ambiente – CNR, Rapporto Scientifico CNR-ITBM 2/94. Roma, 1994, pp. l, 964. ISBN 88-86096-09-7.

UBA – Umweltbundesamt. Die Umwelt-CD-UMPLIS. Umweltbundesamt, Berlin, I-1996, CD-ROM Edition, Benutzerhandbuch pp. 110, Umweltklassifikation, 1993, pp. iii, 12. (Includes the contents of: UBA, Umweltbundesamt. Umwelt-Thesaurus und Umwelt Klassifikation. Umweltbundesamt, Berlin, 1994, pp. v, 11, 347, 495, 150, 133, 9, total 1145.)

UNEP – INFOTERRA, 1997. *EnVoc — Multilingual Thesaurus of Environmental Terms.* Nairobi: UNEP, pp. xix, 248.

UNEP – INFOTERRA, 1990. *Thesaurus of Environmental Terms.* Nairobi: UNEP, pp. xi, 190.

UNEP – INFOTERRA, 1984. *Thesaurus of Environmental Terms.* Nairobi: UNEP, pp. ii, 66.

Vilhena, D. A., Foster, J. G., Rosvall, M., West, J. D., Evans, J., and Bergstrom, C. T., 2014. Finding cultural holes: How structure and culture diverge in networks of scholarly communication. *Sociological Science*, 1, pp. 221–238.

4 Discourses of environmental protection

An ontology approach to domain modelling for translation and multilingual text production

Adriana S. Pagano, André L. Rosa Teixeira, and Davi Seabra Grossi

4.1 Introduction

The globalization of environmental issues has evolved discourses mobilizing concepts that are appropriated, transformed and sometimes contested through meanings construed within and across languages (Tsing, 1997, 2000). A major role in the generation and propagation of such discourses is played by international organizations that produce knowledge and support publications aimed to promote, regulate and report on conservation efforts. International organizations frequently use one language as the main language of communication and have a policy of translation and/or multilingual text production covering a limited choice of other languages. Translated or multilingual texts are produced, not only for the purpose of providing content accessibility, but also, and most importantly, for inducing and aiding local adoption of principles and guidelines towards common transnational goals.

In such contexts, translation is an intrinsic process to organizational procedures and understanding its *modus operandi* demands understanding the institutional setting within which it takes place (cf. Koskinen, 2011), that is, *who* commissions *who* to translate *what* for *whom*. Likewise, in order to understand how transnational discourses get translated by and have an impact on national institutions, it is important to examine which meanings are construed to name new constructs in distinct cultures in response to emerging international concerns. Meanings as construed and worded by each culture can be mapped in an attempt to find translation equivalents operating for particular types and purposes of translation. These two outlooks on institutional translation, – the source institution's and the target institution's – allow us to explain some of the complexity involved in translating global discourses, an operation which, as we will attempt to show, challenges conventional terminological approaches.

The discourse of environmental protection is a case in point, which this chapter explores focusing on conservation efforts in Brazil as a case study illustrating the need for a comprehensive approach to the translation of domain terminology. Bearing upon the assumption that domain terminology implicates rich taxonomies where concepts are interrelated, an ontology approach is here proposed, whereby terms naming concepts in different contexts of culture can be organized with sufficient detail on their intrinsic relations to other terms in the domain so that mappings of equivalent instances can be made and retrieved. In this particular case, we focus on an international organization having English as its main language for knowledge production and communication – the International Union for the Conservation of Nature (IUCN) – and Brazilian institutions devoted to environmental protection education, policy making and implementation.

The analysis leading to the designed ontology revealed the underlying relations in which the concepts are implicated and its output can be used to aid the comprehension of non-specialists participating in policy-making and reporting of ecological conservation matters as well as translators and language professionals dealing with tasks requiring knowledge of that particular domain.

The following sections in this chapter will first present a brief overview of the theoretical framework drawn upon in order to carry out our study, followed by the methodology adopted and results. The concluding section will point out how institutional translation can benefit from an ontology perspective on terminology.

4.2 Literature review

4.2.1 *Institutional translation*

The role of institutions as regulators of translation activities has been explored since the consolidation of the discipline of translation studies in the 1990s (see Mossop, 1990), especially in descriptive approaches examining norms (Toury, 1999) and systems operating in translation (Hermans, 1997). More recently, studies on the topic have been subsumed under the proposal of a sociology of translation (Wolf, 2007), which investigates aspects of the social context of translation bearing an impact on translation practices. Institutions, in particular, have received special attention and have been the object of a dedicated topic in the field, namely "Institutional Translation" (Koskinen, 2011). Regarding this topic, several studies have been published on governmental and non-governmental institutions, focusing on their language and translation policies, both in the legal and other domains (cf. Tesseur, 2014; 2017; Ellis, 2017; Prieto-Ramos, 2017). These studies bring to light factors at play particularly in transnational institutions involved in global policy making and governance, and contribute new angles on the analysis of actors, processes and products of translation and multilingual text production (Koskinen, 2008).

Interestingly, findings of studies on translation in transnational institutions have been shown to challenge established assumptions in the discipline. This is the case of the purposes for which translation is undertaken. Koskinen (2000:51), for instance, shows that in institutions like the European Union translations into languages as Finnish are produced solely for the sake of their being "a proof of linguistic equality" to available translations in other languages, with no real communicative need for them.

Research carried out on institutional translation has focused on the source institutions that produce source texts, commission translations and set up international guidelines for policy making and governance. There is, however, a further aspect to be explored. This has to do with the fact that the meanings construed by source institutions, worded in a source language and reworded in target languages through the translations they commission, are negotiated by target institutions which adopt and/or locally adapt those wordings. In so doing, variation is generated, not only in the translation relation between source and target texts, but also in the source and target language systems. A systemic-functional approach to translation, as we will show next, can illuminate this type of variation.

4.2.2 Social and semiotic institutions

As a comprehensive theory of language, systemic-functional linguistics (henceforth SFL) (Halliday & Matthiessen, 2014) offers a powerful framework to model translation and multilingual text production (Halliday, 1992). This is due to its view of language as social semiotics, that is, a semiotic system embedded in a social system. The semiotic system is, in turn, conceived of as the system of language embedded in the system of context. Figure 4.1 shows a representation of SFL's model.

As Figure 4.1 shows, language is a semiotic system embedded in a social system. As a system, it encompasses the meaning potential of a culture. This potential is deployed along a cline from full potential to instance, there being a mid-point along the cline where subsystems of meaning can be located. These subsystems can be approached as subsystems of social organization which have counterpart subsystems of meanings in the semiotic system. Institutions can thus be viewed as subsystems both of the social and the semiotic systems.

In SFL theory (Matthiessen, Teruya & Lam, 2010), institutions are interpreted as subsystems of meaning (semiotic institutions) that integrate subsystems of

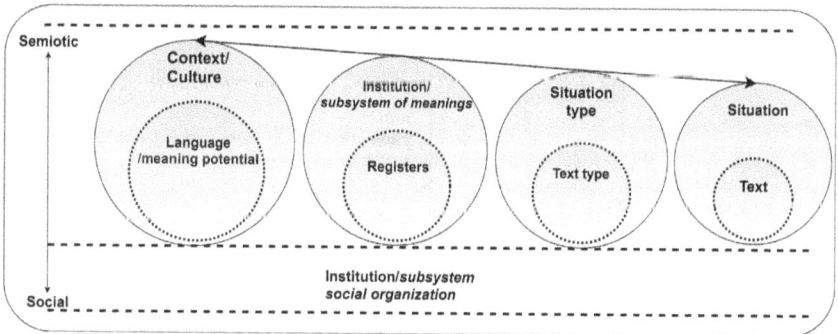

Figure 4.1 Location of institutions as subsystems of social organization and meanings within a SFL framework.

social organization (social institutions). From the perspective of language, institutions can be studied by looking at the range of registers and text types operating within them.

Subsystems at the mid-point in the cline between system and instance are analysed contextually by means of parameters that are relevant for the production of registers and text types. These parameters are subsumed under the concept of context of situation, which encompasses variables for field (domain variables), tenor (interpersonal variables) and mode (text variables), as shown in Figure 4.2.

The variables in Figure 4.2 characterize language-in-context and apply to source and target text production, each of them operating within its own context of culture. Thus, a source text can be analysed in terms of each subvariable, as illustrated in Table 4.1, for one of the texts examined in our case

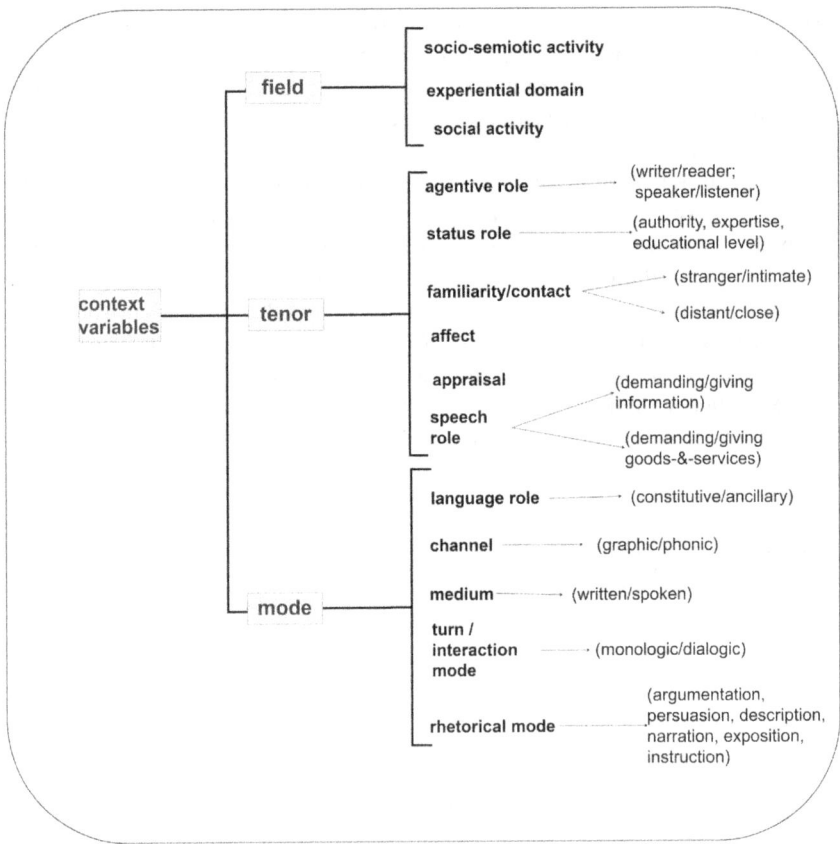

Figure 4.2 Variables and subvariables for modelling context with examples of observable values.

Table 4.1 Analysis of contextual variables for IUCN guidelines on protected areas

Variable	Subvariable	Observation
Field	Socio-semiotic activity	To introduce concepts within a taxonomy (expounding knowledge on protected areas)
		To instruct readers on how to apply knowledge (enabling use of categories of protected areas)
		To regulate behaviour (controlling use of categories of protected areas)
	Experiential domain	Environmental conservation
	Social activity	Institutional guidelines at international level to expound, regulate and instruct at local level
Tenor	Agentive role	Writer: IUCN International institution of environmental experts
		Reader: local environmental institutions aiming to implement international guidelines
	Social role (authority, expertise, educational level, etc)	IUCN occupies authority, and higher expertise roles
	Social distance – familiarity /contact (stranger/intimate; distant/close)	Distant
	Affect	Seeking cooperation, compliance
	Appraisal	Positive evaluation (judgement and appreciation)
	Speech role (demanding/giving information; demanding/ giving goods & services)	Giving information and goods-&-services (mainly the latter)
Mode	Language role (constitutive/ ancillary)	Constitutive
	Channel (graphic / phonic)	Graphic
	Medium (written/spoken)	Written
	Turn / interaction mode (monologic / dialogic)	Monologic
	Rhetorical mode (argumentation, persuasion, description, narration, exposition, instruction)	Exposition of categories; instructions to apply them

study, IUCN *Guidelines for Applying Protected Area Management Categories* (Dudley, 2008).

In the case of translation, as we will see next, text production operates conditioned by variables of what can be posited as a context of situation for the translation process itself or (meta)context of translation. This can be characterized by variables of field, tenor and mode that have an impact on the decisions made by the translator, in turn, impacting her/his language choices for the target text.

4.2.3 *The (meta)context of translation*

Drawing on systemic-functional theory, Matthiessen (2001) proposes to approach translation as operating in environments. These environments are the locus where language choices are made conditioned by different dimensions in language architecture, among them strata (phonology, lexicogrammar, semantics), ranks (morpheme, word, group, clause) and metafunctions (systems of ideational, interpersonal and textual meanings).[1] A further conditioning, still, is the environment of translation as a social practice. Matthiessen (2001) proposes to call this environment "(meta)context of translation" and to characterize it by means of the variables of field, tenor and mode following the set of subvariables presented in Figure 4.2.

Regarding the (meta)context of translation, field has to do with the particular process that is envisaged within various multilingual possibilities (whether the prospective translation is being requested as gloss translation; full translation; partial translation) and sequences of processes that may be part of the translation assignment (pre-editing; post-editing; revision), including modes of human- or machine-assistance. Tenor is related with the different roles in the translation process (who commissions, who performs, who evaluates the translated text); power relations and expertise; targeted audiences, among other aspects. Mode covers aspects pertaining to how the translated text will be produced: as a written or spoken text; in electronic format or not; with computer assistance or not; whether the translated text will operate as if it were a source text in the target culture or as an aid for target readers to access the source text; and the purpose for what the translation will be carried out (for example, to be part of an archive; to be used as a source for information; to fulfil a legal requirement, among many others).

Table 4.2 shows an analysis of the (meta)context of translation with subvariables that can be abstracted from an analysis of IUCN translation policy.

Variables in the (meta)context of translation are a source of variation found in the relation between source and target texts, as different target texts of the same source text can be produced, depending on the combined choices made for each of the (meta)contextual variables. Moreover, as target texts are text instances related through the cline of instantiation to the language system, variation in texts leads to variation in the system. This is one of the reasons why translation per se is a process that impacts languages in contact. Texts instanced in the target language as a product of translation and multilingual processes are part of the target language system.

Variation pertaining to specific domains is focused on in terminological studies. Following Matthiessen's modelling of translation within a more abstract environment, that of the (meta)context of translation, terminology is the output of a series of contextualizations impacted by both the contextual and the (meta)contextual variables of text production. This will become clear in the results section of our case study.

Table 4.2 Analysis of (meta)contextual variables of the translation of IUCN guidelines on protected areas

Variable	Subvariable	Observation
Field	Socio-semiotic activity	To enable comprehension by target readers not proficient in English
	Experiential domain	Environmental conservation
	Social activity	Full translation of guidelines for content accessibility in local language
		Human translation
Tenor	Agentive role	Commissioner of translation: iucn translator: translation services hired by iucn
		Target audience: non-english speaking institutions
	Social role (authority, expertise, educational level, etc)	IUCN occupies authority, and higher expertise roles in environmental policies
		Translator occupies authority and higher expertise in translation activity
	Social distance – familiarity /contact (stranger/intimate; distant/close)	Distant
	Affect	Seeking cooperation, compliance
	Appraisal	In line with source text; no contestation
	Speech role (demanding/giving information; demanding/giving goods & services)	IUCN demanding goods-&-services (commissioner of translation); Translators giving goods-&-services (translation performers)
Mode	Language role (constitutive/ ancillary)	Ancillary
	Channel (graphic / phonic)	Graphic
	Medium (written/spoken)	Written
	Turn/ interaction mode (monologic / dialogic)	Monologic
	Rhetorical mode (argumentation, persuasion, description, narration, exposition, instruction)	Exposition

4.2.4 Terminological variation

Language variation observed in a translation relation between source and target texts is generally referred to as interlingual variation, whereas variation within the source and target language systems is usually discussed as intralingual variation.[2]

The topic of intra- and inter-lingual variation has been object of research in translation studies and terminology studies. The latter, in particular, have drawn on corpus studies and explored resources to organize variation in databases which translators can consult to solve translation problems. León-Araúz (2017), for instance, reports on two projects approaching terminology variation from a multidimensional perspective, one of them in the health care domain (Varimed[3]) and another one on the environmental domain (Ecolexicon[4]). The projects make use of rich annotation of terms and corpus search in order to identify semantic relations, esp. hyperonyms, meronyms and non-hierarchical relations. Results are stored in terminology databanks, there being a web interface available for queries to assist text production and translation. In another study, Kerremans (2017) explores variation extracted from a multilingual corpus of parallel texts on environmental matters to propose a resource to be integrated into a translation tool.

Variation certainly poses a challenge to terminology approaches to translation tasks in that criteria for choice of terms cannot be easily solved by relying on corpus querying, thus requiring domain expert assistance. A proposed solution to such problems is to approach terminology from an ontology perspective, whereby terms are analysed as concepts and organized in taxonomies that make up a domain. Concepts enter into thick relations that are taken for granted by domain specialists, but are obscure to non-specialists and laymen. Ontologies emerge then as semantic representations that enable querying to assist translation and multilingual text production and which can be reused and augmented as the domain expands. In that respect, ontologies are more resourceful than dictionaries, thesauri and conventional terminological banks.

4.2.5 Ontologies as semantic representations

A widely acknowledge definition of an ontology is Gruber's (1993) and Borst's (1997). Gómez-Pérez and colleagues (2004) outline the development of the concept by showing contributions by Studer and colleagues (1998), that merge both definitions and, explain in detail that:

> An ontology is a formal, explicit specification of a shared conceptualization. Conceptualization refers to an abstract model of some phenomenon in the world by having identified the relevant concepts of that phenomenon. Explicit means that the type of concepts used, and the constraints on their use are explicitly defined. Formal refers to the fact that the ontology should be machine-readable. Shared reflects the notion that an ontology captures consensual knowledge, that is, it is not private of some individual, but accepted by a group.

The authors go on and show that Uschould and Jasper (1999) provide a more accessible explanation to a wider range of research communities, by pointing out:

> An ontology may take a variety of forms, but it will necessarily include a vocabulary of terms and some specification of their meaning. This includes definitions and an indication of how concepts are inter-related which

collectively impose a structure on the domain and constrain the possible interpretations of terms.

Some key points emerge from the explanations above: an ontology captures the specificity of a domain as it structures and imposes constraints on concepts and their interrelations. An ontology also involves formalization in a language comprehensible by both human and machines. An ontology entails explicitation and shared agreement by a community of users.

As forms of knowledge organization, ontologies are frequently compared to dictionaries and thesauri. Lassila and McGuinness (2001) classify ontologies, glossaries, thesauri, and others, according to the kind of information they need to present and the complexity and richness of their internal structure, ranging from Controlled vocabularies to ontologies that express general logical constraints (Heavy weight ontologies). Differences can be located in terms of how much organization is reliant on natural language; how much semantic information is offered; what kind of relations are provided; whether hierarchies are supplied; and what kind of restrictions are modelled to make meanings more specific. Table 4.3 illustrates those differences with one of the terms examined in our case study "national park".

As Table 4.3 shows, ontologies encompass a higher amount of information regarding a concept and its relations than glossaries or thesauri. While a query for our search expression "National Park" in controlled vocabularies yielded a list of terms related along a hierarchy of terms, in thesauri we can see a more complex result, as the search yields information such as a definition, related terms and synonyms. Our ontology yielded results regarding different types of properties of the class named "National Park". The OWL language used to author the ontology in the software Protégé (Musen, 2015) allows the insertion of metadata for classes, individuals, and object properties, through Annotation properties such as rdfs: comment, rdfs:label definition, has_alternative_id, and so on. The Class Annotation Tab on Protégé shows the different types of annotation implemented for the class "Natural Park" as shown in Table 4.3. In addition to Annotation properties, Protégé also allows Object properties insertions to establish relations between classes, between individuals and between classes and individuals. In Table 4.3, we can see that "Natural Park" is related to "Genetic Resources Maintainance" through the property "has Primary Objective".

Information packed in ontologies is explicit and, even though natural language is used for user interface, ontologies are authored in formal languages, which allows for querying on domain-specific elements (classes, properties, individuals and their relations) and drawing logical inferences through reasoning mechanisms. Besides that, due to their digital format and design in languages following the World Wide Web Consortium xml standards, ontologies can be expanded and reused as well as integrated in larger arquitectures for a myriad of uses, among them, natural language generation, translation and multilingual text production.

The study herein reported focused on a terminological problem pointed out by Brazilian environmental experts, which as our results will show can be modelled by means of an ontology.

Table 4.3 Sample query results in different knowledge organization web resources

Type of structure	Main features	Query result for search expression "national park"
Controlled vocabularies	Finite list of terms, weak semantics, syntactic interoperability, usually a component in an architecture made up of databanks.	Site Type anthropogenic park protected area national park[11]
Glossaries	List of terms with definition in natural language	**National Park**: Protected area managed mainly for ecosystem protection and recreation (IUCN).[12]
Thesaurus	Provides definitions, synonyms (may provide translations, etc) + visual display of terms, syntactic interoperability	**Definition** Areas of outstanding natural beauty, set aside for the conservation of flora,(...) within national parks, as is industrial activity. **Related terms Broader**: protected area **Themes**: environmental policy natural areas, landscape, ecosystems Group: LAND (landscape, geography) **Other relations** Has close match: UMTHES: National park Has exact match: AGROVOC: National parks EuroVoc: national park **Wikipedia article**: National park[13]
Ontologies (OWL-DL)	Based on Description Logic, strong semantics, semantic interoperability, allow automatic reasoning and inference	**Annotation properties**: -rdfs:label: IUCN national park -has_alternative_id IUCN-PACS:II; -has_broad_synonym national park -definition- An IUCN protected area which 1) primarily consists of ecosystems (...), education, tourism, subsistence use by indigenous communities, and recreation. **Relations (object properties)**: -subclassOf 'area protected according to IUCN guidelines' hasPrimaryObjectivesome (GeneticResourcesMaintainance or HabitatsEcosystemsSpeciesPreservation or PhysiographicRegionNaturalStatePreservation)[14]

4.2.6 Protected areas as case study in terminological variation

The object of the study herein presented is "protected areas", a key concept in the discourse of environmental conservation. The main international organization for establishing standards and regulation regarding protected areas is the International Union for Conservation of Nature (IUCN).

IUCN's is a "membership Union" made up by "government and civil society organisations". Its mission is acknowledged as being to provide knowledge and tools to help promote human progress, economic development and nature conservation.[5]

Although discussion of protected areas dates back to the early 20th Century, a standard definition and guidelines for their implementation at international level were published in 1994 by the International Union for Conservation of Nature (IUCN). The definition was updated in 2007 and so were the guidelines for application of the six protected area categories established.

In one of its publications, titled *Speaking a common language* (Bishop, Dudley, Phillips, & Stolten, 2004), IUCN explained the reasons for proposing management categories for protected areas, which were the need to categorize conservation areas by means of a common international system "regardless of nomenclature used by nations or consistent to particular languages" (p. 11). The categories were also presented as a means to avoid use of terms with different objectives from those specified for each category. As problems were detected due to lack of understanding of the guidelines by members non-speakers of the three official languages at IUCN – English, French and Spanish—, the organization made the decision to adopt a policy of translations into languages other than the official ones and sought to involve advice from in-country specialists. Ever since, despite sustained efforts to supply multilingual publications, lack of translations into local languages has been recurrently pointed out as a major problem in implementing management categories in a consistent manner so as to obtain better reports from member organizations.

IUCN' s translation policy includes sponsoring original publications by member countries as well as translations of existing materials. A style manual, publishing guidelines, a definitions glossary and a glossary of translated terms are made available at IUCN's portal to prospective authors.

As regards resources made available by IUCN in the Portuguese language, the 2008 *Guidelines for applying protected area management categories* have not been yet translated into this language. Other publications in IUCN's series *Best Practice Protected Area Guidelines*, however, have been translated into Portuguese and some of them bring an insert with names and description of the six main categories of protected areas rendered in Portuguese.

As regards published translations, the book *Governança de áreas protegidas: da compreensão à ação* (Borrini-Feyerabend et al., 2017), a translation of *Governance of protected areas: from understanding to action* (Borrini-Feyerabend et al., 2013), acknowledges the name of a translator and a translation company in its title page. A disclaimer on the same page reads: "IUCN assumes no responsibility for any errors, omissions or language shifts in this translation. In the case of discrepancies, please check the original source.[6] Another publication, *Áreas Urbanas Protegidas: perfis e diretrizes para melhores práticas* (Trzyna, T., 2017), a translation of *Urban Protected Areas: profiles and best practice guidelines* (Trzyna, T., 2014), acknowledges the same translation company as the 2017 volume and includes the same disclaimer above. Both publications provide the same translation for IUCN management categories into Portuguese, in which the term "protected areas" is rendered as "áreas protegidas" [protected areas] in Portuguese.

Interestingly, there is a 1997 publication by Borrini-Feyerabend available in Portuguese at IUCN's portal, *Manejo participativo de áreas protegidas: adaptando o método ao contexto,* which is presented as a translation of *Collaborative management of protected areas: tailoring the approach to the context* (Borrini-Feyerabend, 1996), published in Quito, Equator. No translator is acknowledged, which may be interpreted to mean that the author of that publication translated the publication herself. In the volume, the term "protected areas" is rendered as "unidades de conservação" (conservation units), a rendition that evidences intralinguistic variation in Portuguese and a translation problem pointed out in the environmental literature by Brazilian domain experts, which motivated our study.

A brief history of Brazil's policy for protected areas allows us to understand the reported translation problem. The year 2000 is an important milestone in Brazil's legislation regarding protected areas; this is when Act no. 9.985 instituting a National System of Conservation Units was passed. Some of the categories for protected area management had already been created through several pieces of legislation as early as the 1930s; however, it was with Act No 9.985 that a national system was consolidated. The Brazilian legislation names management categories "unidades de conservação" (conservation units), which are subsumed under the broad term "áreas protegidas" (protected areas), together with "territórios indígenas" (indigenous territories), and other categories.

Brazilian geographer and environmentalist Claudio Maretti (2012) has consistently argued that IUCN´s term "protected areas" can be interpreted from two complementary perspectives: in a broad and in a narrow sense. According to Maretti, in its broad sense, "protected area" is a term naming a concept that does not implicate explicit nature conservation objectives, even though conservation is assumed as an indirect goal. On the other hand, "protected area" in its narrow sense, the author contends, names a concept for a management category whose main and explicitly stated goal is nature conservation.

Bearing in mind Brazilian legislation, Maretti (2012) argues, then, that the term "protected area" in its narrow sense is an adequate translation equivalent for "unidade de conservação" (conservation unit), a term used in Brazil´s System of Conservation Units (SNUC) to name a category entailing specific and explicit nature conservation objectives. Furthermore, the author points out, even though other terms in the Brazilian legislation (e.g. territórios indígenas [indigenous territories]; áreas de preservação permanente (permanent preservation areas); áreas de proteção de mananciais de água [water source preservation areas]) are used to name categories discussed within the discourse of "unidades de conservação" (conservation units), those categories cannot be considered "protected areas" in the narrow sense, and hence, they are not "unidades de conservação" (conservation units), but "protected areas" in the broad sense of the term.

In order to fulfil the need for a term more aligned with the international terminology in English, yet preserving the specificity of concepts in Brazilian legislation, Maretti (2012) proposes to translated the English "protected areas" in the narrow sense of the term as "unidades de conservação" (conservation units) in Brazilian Portuguese and the Portuguese term "unidades de conservação" (conservation units) as "nature protected areas" in English.

The terminological variation observed in the Brazilian Portuguese renditions for the international term "protected areas" is the object of the study undertaken, which we report in the following sections.

4.3 Methodology

The study herein reported was carried in order to design a form of knowledge representation capable of organizing meanings construed by the terms "protected areas" in English and the terms "áreas protegidas" (protected areas) and "unidades de conservação" (conservation units) in Brazilian Portuguese which could be instrumental to assist translation and multilingual text production.

Our methodology comprised the following steps.

1   Definition of domain, purpose and scope of ontology

The following parameters were defined for the ontology:
Domain: Environmental domain, more specifically protected areas
Aim: Engineering a multilingual ontology, drawing on the International Union for Conservation of Nature (IUCN)'s policy and the National System of Conservation Units (SNUC) in Brazil
Scope: Protected Areas in the narrow sense of the term (including main objectives, complementary objectives, stakeholders)

2   Selection of sources of information for ontology modelling

Ontology modelling requires a source of information for the acquisition phase and conceptualization of the domain knowledge. Generally, text in the form of legal documents and academic publications are used, together with domain expert consultation. To build our ontology, information for the acquisition phase was retrieved from IUCN publications introducing international guidelines for the classification and implementation of the management categories (IUCN, 2000). For the Brazilian context, information was retrieved from different sources: (i) the Brazilian Ministry of Environment[7] website; (ii) from Act No. 9.985/2000, which established the National System for Conservation Units (Brazil, 2000) and related legislation; and (iii) from publications by Brazilian geographer and environmentalist Claudio Maretti (2004, 2012), who has consistently reported on the need to clarify terminology regarding "protected areas" in the interaction IUCN – Brazil.

3.   Organization in conceptual maps

Study sessions were held to discuss how to organize information retrieved from our text sources. The results of those discussions is reported on in the case study section. Lexical items naming concepts and the relations those concepts were observed to establish between one another were retrieved from the texts and entered as propositions within the conceptual map editing software CmapTools (Institute for Human and Machine Cognition, 2000).

Modelling a domain as a conceptual map (see Figure 4.3) is a useful inter-mediate step in ontological domain modelling since it allows a visualization of the domain in terms of its concepts and the relations they establish. A conceptual map operates as a preview to enable a broader grasp of the underlying network of relations making up the domain and guides ontology engineering.

The map drawn using the conceptual map tools (Cmap) was exported in text file format containing the propositions stated for the relations established between concepts. The propositions exported from the conceptual map tool follow the pattern concept/linking phrase/concept (see Table 4.4), which in ontological terms equals class/property/class.

The basic constituents in an ontology are: classes, properties (relations), indi-viduals, and axioms (statements of relations holding between 'classes', between 'individuals' and between 'classes' and 'individuals'). In the ontology implemented in our study, for instance, 'IUCN national park' and 'Promotion Of Use For Recreational Purposes' are 'classes'; 'hasPrimaryObjective' is an 'objectProperty', that is, a 'relation', whereby we can create the 'axiom': 'IUCN national park has Primary Objective some Promotion Of Use For Recreational Purposes'. In nat-ural language it amounts to say that natural parks have as a primary objective the promotion of their use for recreational purposes.

'Individuals' can be ascribed to a 'class', being an 'instance' of it. For example, 'Yosemite National Park' is an 'individual' of the 'class' 'IUCN national park'. By stating that, we can infer that 'Yosemite National Park' inherits the characteristics of the class.

The propositions exported from our conceptual map were pasted onto a spread-sheet in a spreadsheet editor (see Figure 4.4). Propositions were segmented into subject/predicate/object mapped onto concept/linking phrase/concept in order to be entered in our ontology as class/property/class. Lexical items naming a class were retrieved and pasted onto a new sheet in order to organize the concepts hierarchically for modelling basic taxonomic relations in the ontology (class and subclass axioms) and restriction axioms.

To model classes pertaining to the international and the Brazilian contexts and establish compatible relations between them, classes deemed congruent were pasted onto one cell to be implemented as a single class in our ontology. Classes particular to each context were pasted separately.

Congruences between IUCN and Brazilian management categories were established and subclasses of the class "conservation units" were labelled with the number tagging international management categories. For example, "Estação

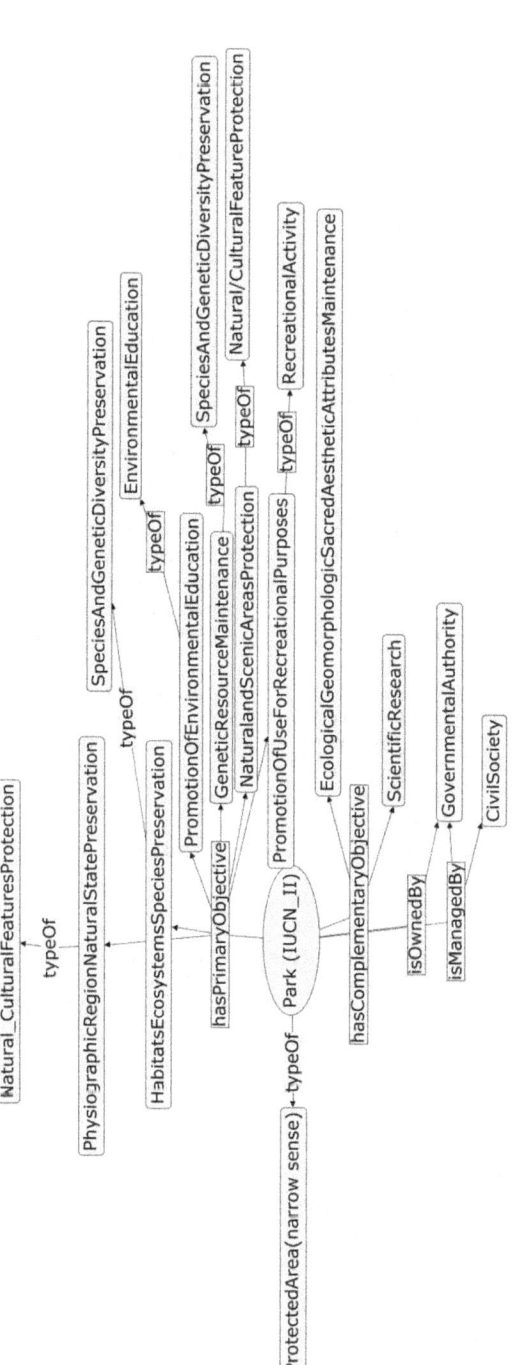

Figure 4.3 Detail of conceptual map for ontology design.

Table 4.4 Sample propositions used to model conceptual map

Concept	Linking phrase	Concept
Natural and Scenic Areas Protection	type_Of	Natural/Cultural Feature Protection
Promotion Of Environmental Education	type_Of	Environmental Education
Physiographic Region Natural State Preservation	type_Of	Natural/Cultural Features Protection
Park (IUCN_II)	is_Owned_By	Governmental Authority
Park (IUCN_II)	has_Primary_ Objective	Promotion Of Use For Recreational Purposes
Park (IUCN_II)	has_Primary_ Objective	Promotion Of Environmental Education
Habitats Ecosystems Species Preservation	type_Of	Species and Genetic Diversity Preservation
Park (IUCN_II)	type_Of	Protected Area (narrow sense)
Park (IUCN_II)	has_Primary_ Objective	Physiographic Region Natural State Preservation
Genetic Resource Maintenance	type_Of	Species and Genetic Diversity Preservation
Park (IUCN_II)	has_Primary_ Objective	Habitats Ecosystems Species Preservation
Park (IUCN_II)	has_Primary_ Objective	Natural and Scenic Areas Protection
Park (IUCN_II)	has_Primary_ Objective	Genetic Resource Maintenance
Park (IUCN_II)	has_Complementary_ Objective	Scientific Research
Promotion Of Use For Recreational Purposes	type_Of	Recreational Activity
Park (IUCN_II)	is_Managed_By	Civil Society
Park (IUCN_II)	has_Complementary_ Objective	Ecological Geomorphologic Sacred Aesthetic Attributes Maintenance
Park (IUCN_II)	is_Managed_By	Governmental Authority

Ecológica" [ecological station] was labelled as "Ia" indicating congruence with IUCN's category "Strict Nature Reserve".

As already said, our ontology was built using the software Protégé. Bearing in mind the principle of reusing existing ontologies, we surveyed ontology portals and found Envo – Ontology of environmental features and habitats (Buttigieg, 2013) – which includes the class "protected areas", informed by IUCN guidelines. That class was imported in our ontology and further specified following the aims of our study.

Subject	Predicate	Object	CLASSES Protected Areas (management categories) strictu	RequiredComponents
NaturalMonument	hasAsObjective	Provision of opportunities for research, education, interpretation and public appreciation	NationalPark	EcosystemProtection
NaturalMonument	is	Area containing one, or more, specific natural or natural/cultural feature which is of outstanding or unique value because of its inherent rarity, representative or aesthetic qualities or cultural significance	WildernessArea	RecreationObligatory
WildernessArea	hasSelectionCriteria	Potential to retain natural attributes if managed properly	StrictNatureReserve	
NationalPark	has	Visitor management and recreation programme	HabitatSpeciesManagementArea	Provision of a foundation for spiritual, scientific, educational, recreational and visitor opportunities, all of which must be environmentally and culturally compatible

Figure 4.4 Screenshot of spreadsheet used to model ontology classes.

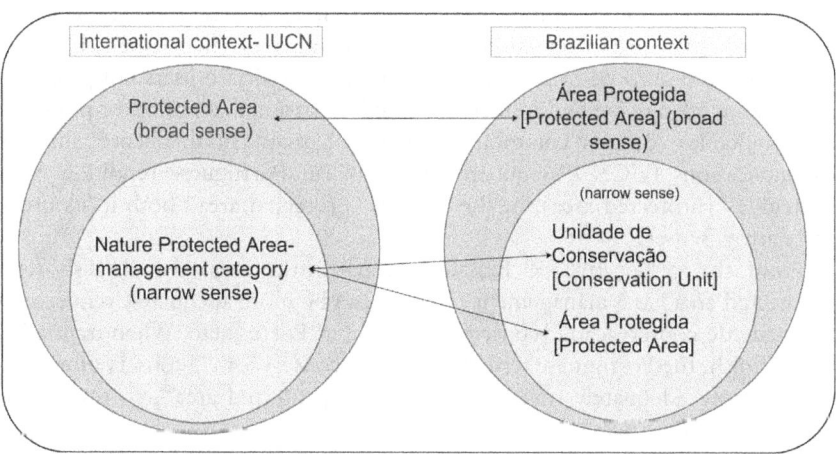

Figure 4.5 Comparison of meaning construed as "protected area" by IUCN and Brazilian legislation.

Hierarchies were manually implemented within Protégé as classes (and subclasses); predicates were modelled as object properties. Restriction axioms between classes were also implemented manually.

Ontology engineering is an iterative activity, that is, it allows researchers to cyclically resort to information sources and implement new classes and properties, whenever there are new findings. In the case of our ontology, once its basic organization had been established, classes which were subsequently considered important (and subclasses and/or their reorganization) were implemented directly on Protégé.

The following section presents the results of our case analysis discussions leading to the conceptual maps underlying our ontology.

4.4 Case study analysis and results

Following Maretti (2012), our main source for comparing the international and the Brazilian contexts regarding protected areas, IUCN's "protected area" in the broad sense of the term names a concept that does not imply explicit nature conservation objectives, despite the fact that conservation is an indirect goal of it. In the narrow sense of the term, "protected area" names a concept for a management category whose main and explicitly stated goal is nature conservation.

These two senses need to be kept separate when finding translation equivalents in Brazilian Portuguese. As Maretti (2012) points out, "protected areas" in the sense of management category with explicit goals is equivalent to "unidades de conservação" [conservation units] within Brazil's legislation on environmental conservation. Figure 4.5 shows a mapping between "protected area" in its broad and narrow senses and its counterparts in the Brazilian context.

From the perspective of the source institution, in this case IUCN, translation and multilingual text production within the context of environmental protection serves the purpose of providing content accessibility to users not proficient in the source language and inducing and aiding national adoption of principles and guidelines towards common transnational goals. In this sense, through its publications IUCN disseminates the Brazilian Portuguese rendition "área protegida" [protected area] for the English "protected area" both in its broad and narrow senses.

From the perspective of Brazilian target institutions, the English term "protected area" as a management category can be more adequately rendered as "unidade de conservação" [conservation unit] in Portuguese. When translating into English, the Portuguese term "unidade de conservação" [conservation unit] can be more adequately translated as "nature protected area", to restrict the broad term "protected area".

Focusing on "protected areas" in the narrow sense of this term, IUCN adopts seven management categories, which are labelled and numbered thus: Strict Nature Reserve (Ia), Wilderness Area (Ib), National Park (II), Natural Monument (III), Habitat/Species Management Area (IV), Protected Landscape/Seascape (V), and Managed Resource Protected Area (VI). These are shown in Figure 4.6.

Through the National System of Conservation Units (SNUC), the Brazilian legislation distinguishes two types of Conservation Units. "Full Protection Units" comprise "Biological Reserve", "Ecological Station", "Park", "Natural Monument", and "Wildlife Refuge". "Sustainable Use Units" comprise "Private Reserve of Natural Heritage", "Area of Relevant Ecological

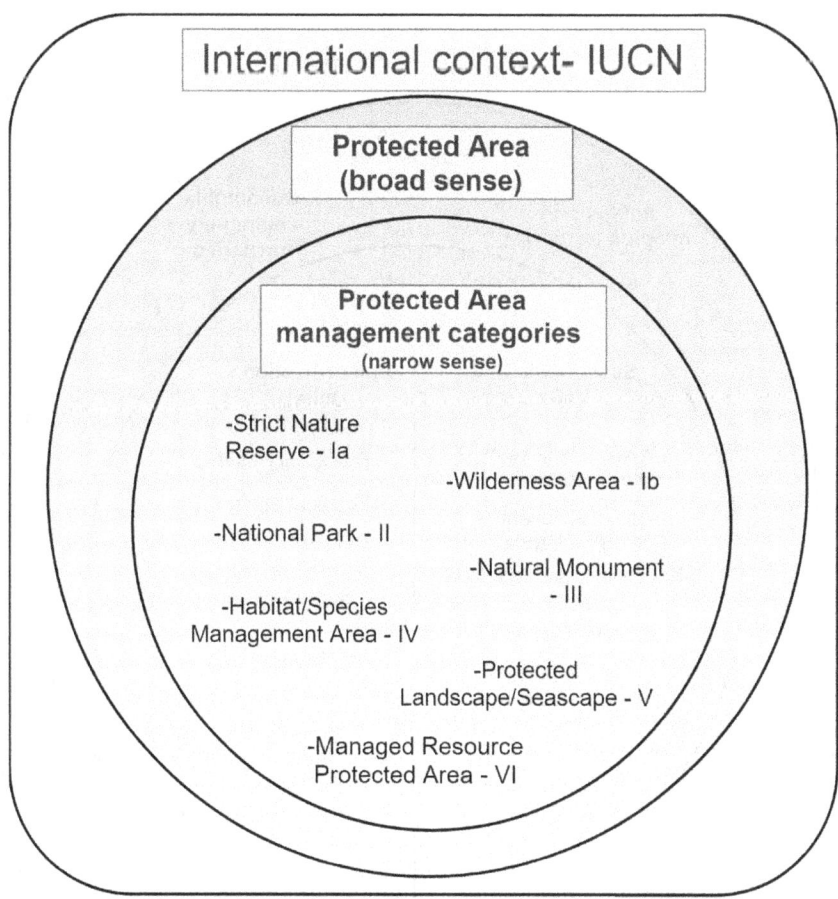

Figure 4.6 IUCN's management categories of protected areas.

Interest", "Environmental Protection Area", "Extractive Reserve", "Sustainable Development Reserve", "Forest", and "Fauna Reserve". Within the broader scope of "protected areas" are, for instance, "Indigenous peoples' Territories", "Quilombola community Territories", with specific management goals and criteria different from those of "conservation units" or "protected areas" in the narrow sense of the term.[8] This is shown in Figure 4.7.

Maretti (2012) correlates Conservation Units in Brazilian context with IUCN's management categories. Figure 4.8 shows a mapping between Brazilian Full Protection Units and IUCN's management categories. Biological Reserve and Ecological Station are congruent with Strict Nature Reserve (Ia); the Brazilian category Park[9] is congruent with National Park (II); and Brazilian Natural Monument is congruent with IUCN Natural Monument (III). The

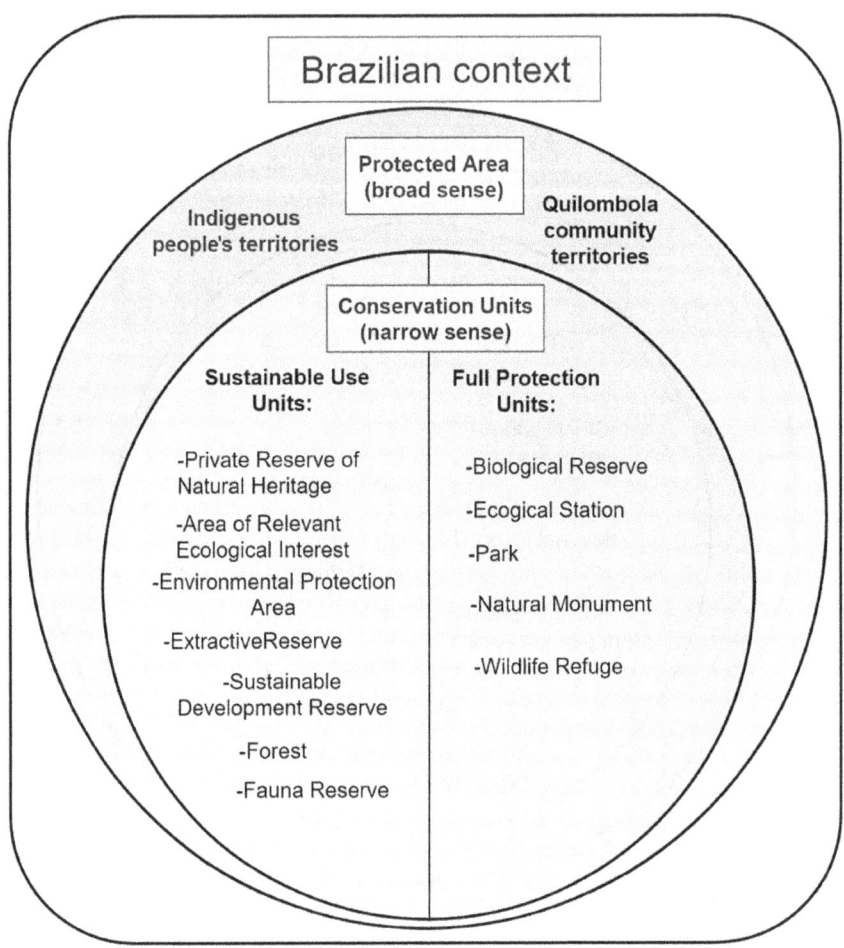

Figure 4.7 Conservation units – Brazilian legislation.

Brazilian category Wildlife Refuge does not have a full equivalent in IUCN's Protected Area management categories.

Figure 4.9 shows a mapping between Brazilian "Sustainable Use Units" and IUCN's management categories. In this case, congruence is established between "Private Reserve of Natural Heritage" and "National Park" (II); "Area of Relevant Ecological Interest" and "Natural Monument" (III); "Environmental Protection Area" and "Protected Landscape/Seascape" (V); "Extractive Reserve", "Sustainable Development Reserve", "Forest", and Fauna Reserve" and "Managed Resource Protected Area" (VI).

Relations established between IUCN's management categories and the whole spectrum of Brazilian "conservation units" are shown in Figure 4.10.

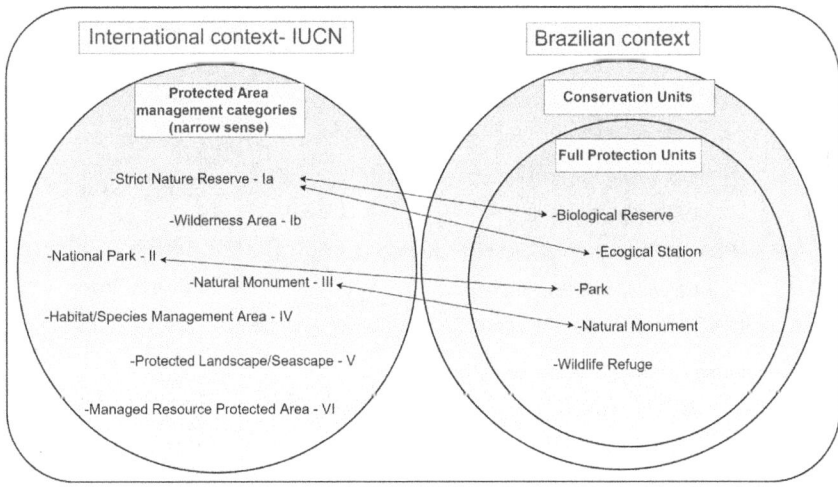

Figure 4.8 Mapping between Brazilian "Full Protection Units" and IUCN' s management categories.

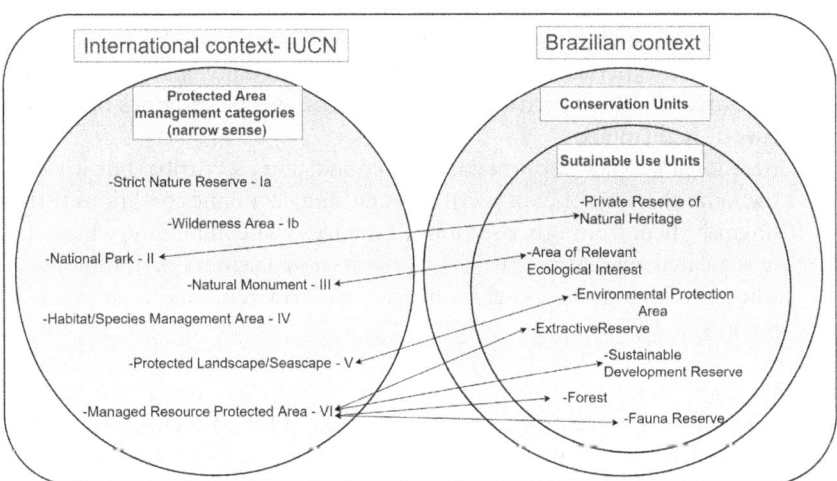

Figure 4.9 Mapping between Brazilian "Sustainable Use Units" and IUCN's management categories.

A further source of translation problems due to terminological variation is found for the term "indigenous people" as a category in IUCN guidelines. IUCN's publication *Standard on Indigenous Peoples* (2016) acknowledges the International Labour Organisation (ILO) Convention on *Indigenous and Tribal Peoples in Independent Countries* as a source for IUCN's definition of "indigenous peoples", which reads thus:

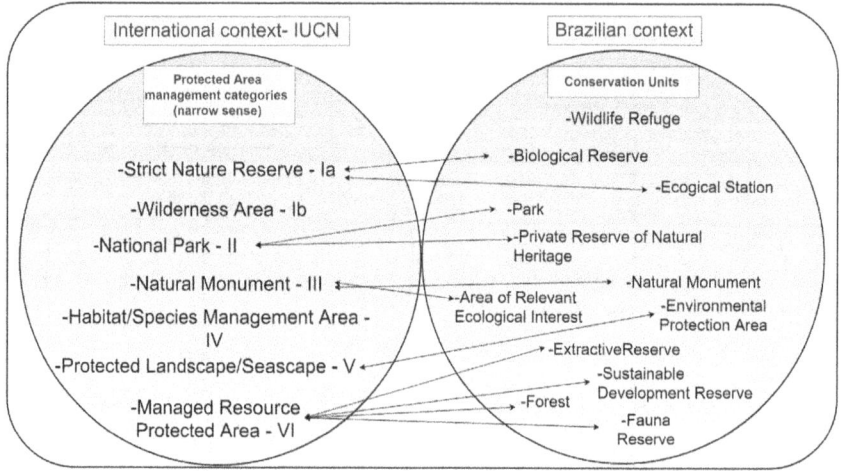

Figure 4.10 Mapping between Brazilian "conservation units" and IUCN's management categories.

i   peoples who identify themselves as 'indigenous';
ii tribal peoples whose social, cultural, and economic conditions distinguish them from other sections of the national community, and whose status is regulated wholly or partially by their own customs or traditions or by special laws or regulations;
iii traditional peoples not necessarily called indigenous or tribal but who share the same characteristics of social, cultural, and economic conditions that distinguish them from other sections of the national community, whose status is regulated wholly or partially by their own customs or traditions, and whose livelihoods are closely connected to ecosystems and their goods and services. p.2.

As the definition reveals, "indigenous people" is therefore used in a broad sense of the term, comprising three groups of peoples. One of them is called "indigenous peoples" in the narrow sense of the term, the other two being called "tribal peoples" and "traditional peoples".

In addition to self-identification, mentioned in IUCN guidelines, the ILO Convention defines "indigenous" in terms of (i) descendancy from populations that inhabited a land prior to colonization by other peoples or to foundation of a present State and (ii) preservation of some or all prior cultural institutions.

In Brazil, the 1988 Constitution acknowledges the rights of "indígenas" (indigenous peoples) and "quilombo" peoples to their land and culture preservation. The former refers to indigenous peoples in the narrow sense of the term whereas the latter to descendants of former slaves. The so-called "traditional populations" are not explicitly mentioned in the Constitution, but their rights are acknowledged in subsequent legislation. Federal Decree 6.040 (2007) defines traditional peoples and communities and traditional territories, highlighting their right to self-identification and the inherent relation between their livelihoods and the territories and natural resources they occupy and make use of, both on a permanent or a temporary regime. Among the many identity categories with which traditional communities identify themselves are "sertanejos" (sertão dwellers), "seringueiros" (rubber tappers), "comunidades de fundo de pasto" (collective pasture communities in the State of Bahia), "agroextrativistas" (Amazonian agroextractivists), "faxinais" (rural inhabitants of the state of Paraná), "pesca artesanal" (artisanal fishing communities), "ciganos" (Roma peoples), "pomeranos" (pomeranians), "pantaneiros" (pantanal dwellers), coconut breakers, "caiçaras" (rural coastal populations), "gerizeriros" (inhabitants of the cerrado/Brazilian savannah), "castanheiros" (Brazilian-nut collectors); berbigão (shellfish collectors); "babaçueiras" (babaçu extractive-related people), "ribeirinhos" (river dwellers), "andirobeiras" (andiroba extractors). "apanhadores de sempre-vivas" (dry-flower collectors), "vazanteiros" (lowlands dwellers), "cipozeiros" (vine gatherers).

Decree 6.872 (2009) establishes the National Plan for Racial Equality Promotion Policies and distinguishes three main groups as targets of affirmative actions: indigenous peoples, quilombola communities and terreiro[10] communities, later renamed as "Communities of African origin". The latter became the target of public policies for preservation of Afro-Brazilian religions and thus acquired a status apart from traditional communities established in Decree 6.040 (2007).

A comparison of IUCN's and Brazil's definition of indigenous peoples is shown in Figure 4.11.

Figure 4.11 reveals a clear distinction between IUCN and the Brazilian context. When translating the English term "indigenous peoples" in the broad sense of the term, the most adequately translation into Brazilian Portuguese is "povos indígenas e comunidades tradicionais" (indigenous peoples and traditional communities). The English term "indigenous peoples" in the narrow sense of the term is equivalent to "povos indígenas" in the Brazilian context. To translate the Brazilian term "povos indígenas" (indigenous peoples) into English demands a further specification of IUCN's broad definition so as to make it clear that "traditional communities" are not encompassed by the term in Brazilian legislation.

Constitutional law has an impact on environmental legislation and policy in Brazil, which likewise establishes a distinction between "indigenous peoples" and "traditional populations". When it comes to "protected areas", indigenous peoples' rights to their land are granted regardless of conservation purposes and there is thus a divide between "indigenous territories" and "conservation units" or "protected areas" in the narrow sense of the term. In the case of "traditional peoples", recognition of their rights is negotiated in terms of their history of preservation of cultural values and practices that involve community living and service to the community, together with environmental services and low environmental impact. Hence, "traditional peoples" are in fact implicated in some of the categories of "conservation units" for sustainable use, as is the case of "Extractive Reserve" and "Sustainable Development Reserve".

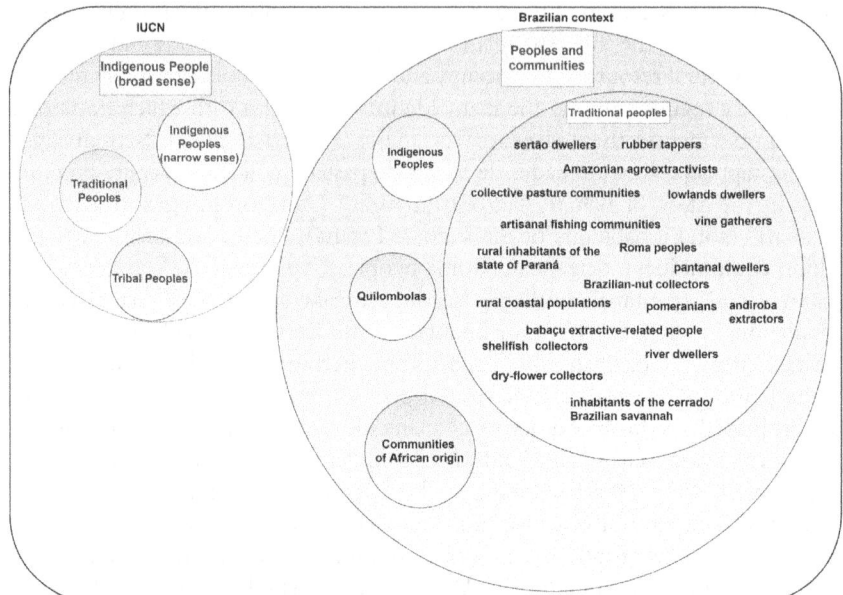

Figure 4.11 IUCN's and Brazil's definition of indigenous peoples.

The legally established dividing line between "indigenous territories" and "conservation units" or "protected areas" in the narrow sense of the term is actually a source of dispute as there are indigenous communities inhabiting lands within conservation units. There is yet no legal provision for overlapping between the two categories despite territory overlapping.

For the purposes of domain modelling, the comparison between the international and the Brazilian views on "protected areas" reveals a series of features that restrict the meaning construed by each category. In order to model these restrictions, a more comprehensive form of knowledge organization is needed than that of a glossary or thesaurus, one that can account for the different terms in their specific contexts of use with the restrictions that apply in each case.

As stated, the perspective adopted by the source institution for international nomenclature, in this case IUCN, is that of translation and multilingual text production for the purpose of providing content accessibility to users not proficient in the source language. This is why, through its publications, IUCN disseminates the Portuguese rendition "área protegida" (protected area) for the English "protected area" both in its broad and narrow senses. From the perspective of Brazilian target institutions, as we have repeatedly pointed out in our study, "protected area" as a management category in the strict sense of the term can be more adequately rendered as "unidade de conservação" (conservation unit). Similar restrictions are observed for the translation of the "erm indigenous peoples" as discussed above.

Restrictions relevant to terms for the purpose of translation can be made explicit by means of propositions to be entered in an ontology as axioms. In the case of "protected areas", some of the many axioms as retrieved from Protégé are illustrated in Table 4.5.

A graph visualization of some of the axioms in Table 4.5 can be seen in Figure 4.12.

As a form of explaining the intrinsic semantic network underlying terms and the concepts they name, an ontology approach to domain knowledge organization is a powerful resource for building and sharing domain knowledge not only for target institutions to translate and accommodate an ever expanding stock of terms, but also for international institutions to better improve their translation policies.

4.5 Conclusion

International institutions, such as IUCN, play a major role in the production, and dissemination of content, whereby they envisage to promote and regulate environmental protection initiatives internationally. In order to do so, those institution have a policy for translation and/or multilingual production in order to provide content accessibility or aid local adoption of international guidelines. In this context, understanding the institutional setting for the translation plays a major role if one is to understand the complexity involved in the process. Thus, it is important to understand both the source institution's and the target institution's outlooks on institutional translation. we need to understand the impact of local specificities on meaning construal, to map contexts and find suitable translation equivalents.

Our study focused on terminological issues pointed out by environmental experts in the Brazilian context, particularly the key concept in environmental conservation "protected areas". We have shown that, in general, studies on institutional translation focus on the source institution, but that there is a further aspect to be explored, that focuses on the terminological variation generated by the context of translation itself. Variation poses challenge to terminology approaches, as results yielded by corpus queries may prove unsuitable for translation problem solving. We have argued that a solution to problems posed by traditional terminology approaches is to explore an ontology perspective. Thus, drawing on systemic-functional theory on language, we have explored the impact of (meta)contextual variables of field, tenor, and mode, on the translation of institutional environmental discourse, and approached terminology from an ontology perspective.

Our case study brings a strong argument for the potential of an ontology approach to terminology, as compared to glossaries and controlled vocabularies, in terms of richness of structure, semantic information offered, implementation of restrictions to make meaning more specific. The potential of an ontology approach to terminology resides in that ontologies are implemented in formal languages, that allow for querying, and inferencing through reasoners. Besides, following W3C standards, they allow for expansion, reuse, integration, enabling use in natural language generation, translation and multilingual text production.

Table 4.5 Axioms pertaining to the classes "protected area", "indigenous territory" and subclasses extracted from Protégé

Axiom
ProtectedArea has BroadSense
ProtectedArea has NarrowSense
ProtectedAreaNarrowSense typeOf ProtectedArea
ProtectedAreBroadSense typeOf ProtectedArea
ProtectedAreaBroadSense not (hasPrimaryObjective some NaturePreserve)
ProtectedAreaBroadSense hasIndirectContribution NaturePreserve
IndigenousTerritory typeOf (Brazilian)ProtectedAreaBroadSense
IndigenousTerritory has DelimitedTerritory
IndigenousTerritory is LargePortionOfLand
IndigenousTerritory has SpecialManagementCriteria
IndigenousTerritory not (hasPrimaryObjective some NaturePreserve)
IndigenousTerritory hasPrimaryObjective
IndigenousPeopleCulturalMaintenance-Protection
IndigenousTerritory has IndigenousPeople
IndigenousPeople liveIn IndigenousTerritory
IndigenousPeople need NaturalResources
IndigenousPeople protect NaturalResources
IndigenousPeople has LowDensityPopulations
IndigenousPeople has TraditionalTechniquesOfResourceUse
IndigenousTerritory hasIndirectContribution NaturePreservation
QuilombolaTerritory typeOf (Brazilian)ProtectedAreaBroadSense
QuilombolaTerritory is SmallerTerritory
QuilombolaTerritory has HighDensityPopulation
QuilombolaTerritory hasIndirectContribution NaturePreservation
QuilombolaTerritory has DelimitedTerritory
QuilombolaTerritory has SpectialManagementRegime
ProtectedAreaNarrowSense hasPrimaryObjective NaturePreservation
ConservatioUnit issynonymOfProtectedAreaNarrowSense FullProtectionUnits typeOf ConservatioUnit
WildlifeReguge typeOf FullProtectionUnits
BiologicalReserve typeOf FullProtectionUnits
EcologicalStation typeOf FullProtectionUnits
Park typeOf FullProtectionUnits
NaturalMonument typeOf FullProtectionUnits
SustainableUseUnits typeOf ConservatioUnit
PrivateReserveofNaturalHeritage typeOf SustainableUseUnits
AreaOfRelevantEcologicalInterest typeOf SustainableUseUnits
EnvioronmentalProtectionArea typeOf SustainableUseUnits
ExtractiveReserve typeOf SustainableUseUnits
SustainableDevelopmentReserve typeOf SustainableUseUnits
Forest typeOf SustainableUseUnits
FaunaReserve typeOf SustainableUseUnits
NatureProtectedArea isA ProtectedAreaNarrowSense
NationalPark typeOf NatureProtectedArea
WildernessArea typeOf NatureProtectedArea
StrictNatureReserve typeOf NatureProtectedArea
HabitatSpeciesManagementArea typeOf NatureProtectedArea
ProtectedLandscape/Seascape typeOf NatureProtectedArea
ManagedResourceProtectedArea typeOf NatureProtectedArea
NaturalMonument typeOf NatureProtectedArea

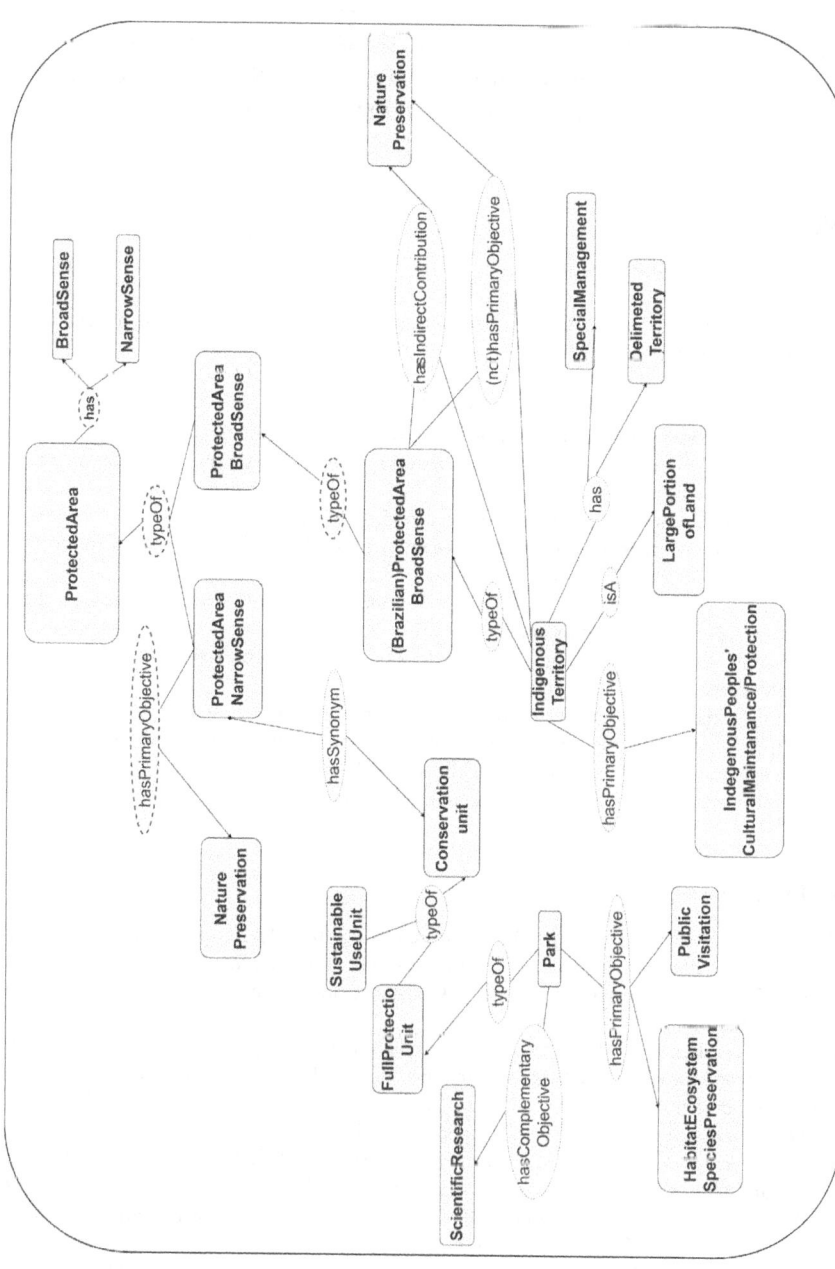

Figure 4.12 Sample graph visualization of ontology classes and subclasses with axioms.

Acknowledgements

Research funded by the National Council for Scientific and Technological Development (CNPQ) under grant No. 310630/2017-7, the Foundation for the Coordination and Improvement of Higher Education Personnel (CAPES) (PhD scholarship) and the State Funding Agency of Minas Gerais (FAPEMIG) under grant APQ-01.461-14.

Notes

1 For a full account of language architecture in SFL theory, see Halliday and Matthiessen (2014).
2 See Steiner (2001) for a full account of intra/interlingual variation and translation.
3 varimed.ugr.es
4 ecolexicon.ugr.es
5 See UICN' s mission statement at www.iucn.org/.
6 A UICN não se responsabiliza por erros, omissões ou desvios da linguagem original da publicação que podem ocorrer nesta tradução. Em caso de discrepâ ncias, por favor consulte a edição original.
7 Available at: www.mma.gov.br/areas-protegidas.html.
8 For a more comprehensive list of "protected areas" in the broad sense of the term in Brazil, see Strapazzon & Mello (2015).
9 In the Brazilian context, Park refers to National Park; State Park, and Municipal Park, under federal, state and local administration, respectively.
10 *Terreiro*, literally yard or grounds, is the term used for the religious spaces used in candomblé.
11 http://edscvs.ccny.cuny.edu/eds/index.php?tema=13989 &/national-p.
12 https://mot.kielikone.fi/mot/endic/netmot.exe?UI=ened&height=165.
13 www.eionet.europa.eu/gemet/en/concept/5484.
14 Data retrieved from ontology implemented in the software Protégé.

References

Bishop, K., Dudley, N., Phillips, A., & Stolten, S. (2004). *Speaking a common language: Uses and performance of the IUCN system of management categories for protected areas.* Cardiff: Union Internationale pour la Conservation de la Nature et de ses Ressources, Switzerland.

Borrini-Feyerabend, G. (1996). *Collaborative management of protected areas: Tailoring the approach to the context.* Gland, Switzerland: IUCN–The World Conservation Union.

Borrini-Feyerabend, G. (1997). *Manejo participativo de áreas protegidas: Adaptando o método ao contexto, Temas de Política Social.* Quito: IUCN-SUR.

Borrini-Feyerabend, G., Dudley, N., Jaeger,T., Lassen, B., Pathak Broome, N., Phillips, A., & Sandwith, T. (2013). *Governance of protected areas: From understanding to action.* Best Practice Protected Area Guidelines Series No. 20. Gland, Switzerland: IUCN, xvi + 124pp. ISBN: 978-2-8317-1608-4.

Borrini-Feyerabend, G., Dudley, N., Jaeger, T., Lassen, B., Pathak, N., Phillips, A., & Sandwith, T. (2017). Governança de áreas protegidas: da compreensão à ação (R. C. Costa, Trans.). Retrieved from https://portals.iucn.org/library/node/46934.

Brazil. Lei Federal Nº 9.985, de 18 de julho de 2000. Regulamenta o art. 225, § 1o, incisos I, II, III e VII da Constituição Federal, institui o Sistema Nacional de Unidades de Conservação da Natureza e dá outras providências., Pub. L. No. 9.985 (2000). Retrieved from www.planalto.gov.br/ccivil_03/leis/L9985.htm.

Buttigieg, P., Morrison, N., Smith, B., Mungall, C. J., Lewis, S. E., & the ENVO Consortium. (2013). The environment ontology: Contextualising biological and bio-medical entities. *Journal of Biomedical Semantics*, 4(1), p. 43. https://doi.org/10.1186/2041-1480-4-43.

Dudley, N. (ed.) (2008). *Guidelines for applying protected area management categories.* Gland, Switzerland: IUCN. https://doi.org/10.2305/IUCN.CH.2008.PAPS.2.en.

Dudley (2013). *IUCN WCPA best practice guidance on recognising protected areas and assigning management categories and governance types, best practice protected area.* Guidelines Series No. 21. Gland, Switzerland: IUCN.

Ellis, J. (2017). The role of translation in transnational governance. *Tilburg Law Review*, 22(1–2), pp. 165–184. https://doi.org/10.1163/22112596-02201008.

Gómez-Pérez, A., Fernández-López, M., & Corcho, O. (2004). *Ontological engineering: With examples from the areas of knowledge management, e-commerce and the Semantic Web.* London; New York: Springer.

Gruber, T. R. (1993). A translation approach to portable ontology specifications. *Knowledge Acquisition*, 5(2), pp. 199–220. https://doi.org/10.1006/knac.1993.1008.

Halliday, M. A. K. (1992). Language Theory and Translation Practice. Campanotto Editore Udine. Retrieved from www.openstarts.units.it/handle/10077/8905.

Halliday, M. A. K., & Matthiessen, C. M. I. M. (2014). *Halliday's introduction to functional grammar* (fourth edition). Milton Park/Abingdon/Oxon: Routledge

Hermans, Theo. (1997). Translation as institution. In M. Snell-Hornby, Z. Jettmarová & K. Kaindl (eds.), *Translation as intercultural communication. Selected papers from the EST Congress Prague 1995.* Amsterdam/Philadelphia: John Benjamins, pp. 3–20

Institute for Human and Machine Cognition. (2000). IHMC CmapTools (Version 6.03) [Windows]. University of West Florida. Retrieved from https://cmap.ihmc.us/cmaptools/cmaptools-download/.

IUCN (ed.). (2000). *Guidelines for protected area management categories: interpretation and application of the protected area management categories in Europe* (2nd corr. edition). Grafenau.

IUCN (ed.). (2016). IUCN Standard on Indigenous Peoples [2.0]. Retrieved May 26, 2019, from www.iucn.org/sites/dev/files/content/documents/iucn_esms_standard_indigenous_peoples.pdf

Kerremans, K. (2017). Towards a resource of semantically and contextually structured term variants and their translations. In P. Drouin, A. Francœur, J. Humbley & A. Picton (eds.), *Multiple perspectives on terminological variation*, Vol. 18. Amsterdam: John Benjamins, pp. 83–108. https://doi.org/10.1075/tlrp.18.04ker.

Koskinen, K. (2000). Institutional illusions: Translating in the EU Commission. *The Translator*, 6(1), pp. 49–65. https://doi.org/10.1080/13556509.2000.10799055.

Koskinen, K. (2008). *Translating institutions: An ethnographic study of EU translation.* Manchester, UK: St. Jerome Pub. and Kinderhook, NY: InTrans Publications.

Koskinen, K. (2011). Institutional translation. In Y. Gambier & L. van Doorslaer (eds.), *Handbook of translation studies*, Vol. 2. Amsterdam: John Benjamins Publishing, pp. 54–60. https://doi.org/10.1075/hts.2.ins1.

Lassila, O., & McGuinness, D. (2001). *The role of frame-based representation on the Semantic web.* Technical Report KSL-01-02. Knowledge Systems Laboratory. Stanford, CA: Stanford University.

León-Araúz, P. (2017). Term and concept variation in specialized knowledge dynamics. In P. Drouin, A. Francœur, J. Humbley, & A. Picton (eds.), *Multiple perspectives on terminological variation*, Vol. 18. Amsterdam: John Benjamins Publishing, pp. 213–258. https://doi.org/10.1075/tlrp.18.09leo.

Maretti, C. C. (2004). Conservação e valores; relações entre áreas protegidas e indígenas: possíveis conflitos e soluções. In R. Fani (series ed.), *Terras indígenas & unidades de conservação da natureza: o desafio das sobreposições.* Retrieved from www.researchgate. net/publication/271842025_Conservacao_e_valores_relacoes_entre_areas_ protegidas_e_indigenas_possiveis_conflitos_e_solucoes.

Maretti, C. C. (2012). Áreas Protegidas: Definições, Tipos e Conjuntos – reflexões conceituais e diretrizes para gestão. In M. O. Cases (series ed.), *Gestão de Unidades de Conservação: compartilhando uma experiência de capacitação.* Retrieved from www. researchgate.net/publication/271842251_Areas_Protegidas_Definicoes_Tipos_e_ Conjuntos_-_reflexoes_conceituais_e_diretrizes_para_gestao/stats.

Matthiessen, C. M. I. M., Teruya, K., & Lam, M. (2010). *Key terms in systemic functional linguistics.* London/New York: Continuum.

Mossop, B. (1990). Translating institutions and "idiomatic" translation. *Meta: Journal des traducteurs,* 35(2), p. 342. https://doi.org/10.7202/003675ar.

Musen, M. A. (2015). The Protégé Project: A look back and a look forward. *AI Matters,* 1(4), pp. 4–12. https://doi.org/10.1145/2757001.2757003.

Prieto-Ramos, F. (2017). Global law as translated text: Mapping institutional legal translation. *Tilburg Law Review,* 22(1–2), pp. 185–214. https://doi.org/10.1163/ 22112596-02201009.

Strapazzon, M. C., & Mello, N. A. D. (2015). Um convite à reflexão sobre a categoria de unidade de conservação refúgio de vida silvestre. *Ambiente e Sociedade,* 18(4), pp. 161– 178. https://doi.org/10.1590/1809-4422ASOC1001V1842015.

Studer, R., Benjamins, V. R., & Fensel, D. (1998). Knowledge engineering: Principles and methods. *Data & Knowledge Engineering,* 25(1), pp. 161–197. https://doi.org/10. 1016/S0169-023X(97)00056-6.

Tesseur, W. (2014). Institutional multilingualism in NGOs: Amnesty International's strategic understanding of multilingualism. *Meta: Journal des traducteurs,* 59(3), p. 557. https://doi.org/10.7202/1028657ar.

Tesseur, W. (2017). Incorporating translation into sociolinguistic research: Translation policy in an international non-governmental organisation. *Journal of Sociolinguistics,* 21(5), pp. 629–649. https://doi.org/10.1111/josl.12245.

Toury, Gideon. (1999). A Handful of Paragraphs on 'Translation' and 'Norms'. In C. Schäffner (ed.), *Translation and norms.* Clevedon/Philadelphia/Toronto/Sydney/ Johannesburg: Multilingual Matters, pp. 9–32.

Trzyna, T. (2014). *Urban protected areas: Profiles and best practice guidelines.* Best Practice Protected Area Guidelines Series No. 22. Gland, Switzerland: IUCN, xiv + 110 pp.

Trzyna, T. (2017). *Áreas Protegidas Urbanas: Perfis e diretrizes para melhores práticas* (R. C. Costa, Trans.). Série Diretrizes para melhores Práticas para Áreas Protegidas No. 22. Gland, Suíça: UICN, xiv + 110 pp. ISBN: 978-2-8317-1859-0.

Tsing, A. (1997). Transitions as translations. In J. W. Scott, C. Kaplan, D. Keates (eds.), *Transitions environments translations: Feminisms in international politics* (first edition). New York: Routledge.

Tsing, A. (2000). The global situation. *Cultural Anthropology,* 15(3), pp. 327–360.

Uschold, M., & Jasper, R. (1999). A framework for understanding and classifying ontology applications. IJCAI-99 Workshop on Ontologies and Problem-Solving Methods (KRR5), Stockholm, Sweden.

Wolf, M. (2007). The emergence of a sociology of translation. In M. Wolf & A. Fukari (eds.), *Constructing a sociology of translation,* Vol. 74. Amsterdam: John Benjamins Publishing, pp. 1–36. https://doi.org/10.1075/btl.74.01wol.

5 Extracting the essence

Toward artificial translation of literature

Mark Seligman

5.1 Introduction

This volume champions sustainability of our world's cultures and languages while exploring the role of translation in that sustenance. Few would question that goal; but just what *is* the relation between sustainability and translation? Several articles herein focus upon dominance relationships among cultures and languages: how can translation help an economically, militarily, or demographically disadvantaged culture to survive in the face of a dominant one? My approach, though, will simply be that translation of any sort – whether real-time or delayed, and whether supplied by humans or by programs – can help to sustain *any* culture in several ways. Most obviously, translation can foster a culture's survival by bridging gaps in in everyday interactions, thus postponing the need to interact in the dominant language. However, my special interest here will be on conservation of a culture through preservation of its literature. In particular, my marching orders are to explore the role of machine translation (MT) and artificial intelligence (AI) in literary translation.

But am I on a suicide mission? This assignment has me walking point, rather far in advance of the main force. Can artificial translation contribute to literary translation at all? After all, artistic translation has long been held up as the quintessential example of what machine translation cannot do now and may never do.

Literary translation strives to somehow preserve the essence of a work while carrying it over to a different language and culture and giving it rebirth there. To recognize that essence, the translator must accurately capture the meaning of the original; appreciate its connotations, register, references, and other abstract or associative factors; and choose among available target language expressions by exercising esthetic judgments. Computers, however, presently remain incapable of such accuracy, abstraction, and judgment.

I'm convinced, however, that MT and other natural language processing, as powered by steadily strengthening AI, will after all become capable of supporting linguistic and cultural preservation. From a practical viewpoint, artificial translation can indeed help preserve whole languages and specific texts as it improves. But advancing MT can also preserve a text in a metaphorical sense by capturing its essence, or essences – at least partially soon, and more as time goes by – to enable its transfer to a new language and culture.

This transfer can be variously imagined. The editors of this volume favor the image of resurrection – of the source text as dead and its translation as a restoral to life (though a literal translation can be conceived as still somewhat dead, while a freer one is more alive). Given that the goal of sustainability is precisely that the things to be preserved – here, texts, languages and cultures – should *not* die, my own preference is for metaphors of keeping alive (though an exception can be made if the source language is itself dead).[1]

Vladimir Nabokov sowed a botanical metaphor in "On Translating Eugene Onegin" (Nabokov, 1955):

> The parasites you were so hard on
> Are pardoned if I have your pardon,
> O Pushkin, for my stratagem.
> I traveled down your secret stem,
> And reached the root, and fed upon it;
> Then, in a language newly learned,
> I grew another stalk and turned
> Your stanza, patterned on a sonnet,
> Into my honest roadside prose –
> All thorn, but cousin to your rose.
>
>   (5–14)

For Nabokov, Pushkin's original Russian rosebush, far from being dead, was immortal. Its roots were thriving and nourished the poet (disguised as a caterpillar, the precursor of a Nabokovian butterfly) as he extracted the bush's essence so as to (somehow!) engender a new shrub which, though thorny, bore an undeniable family resemblance to the first. Nothing died; rather, translation had wrought fruitful multiplication: where one bush had flourished before, now there were two.

The image of translation as extraction and revivification of the original's essence will be helpful throughout this chapter. What is the essence of a text, anyway, and how could an artificial intelligence possibly be brought to recognize and transmogrify it? That question, in fact, constitutes the chapter's own essence.

And Nabokov will aid us in another way, too: the very poem translation which inspired the quoted second-derivative poem occasioned a notorious controversy highlighting several issues to be examined here. Nabokov deliberately produced a hyper-literal translation of Pushkin's verse epic (Pushkin, 1964); the critic Edmund Wilson, Nabokov's closest American friend till then, panned it as lifeless; Nabokov, his honor challenged, counterattacked – and the infamous literary battle was joined, with the friendship an early casualty.

We'll recount this fracas in Section 5.2. Beyond furnishing guilty entertainment, it will illustrate that a text can have multiple aspects that jointly compose its essence, some easier to capture than others. As we go, we'll refer to Douglas

Hofstadter's insufficiently known *tour de force* on translation, *Le Ton Beau de Marot: In Praise of the Music of Language* (Hofstadter, 1997). In it, he translated a short French poem (Clément Marot's "*A une damoyselle malade*") in numerous widely – or wildly – differing styles, thus demonstrating that, if an artwork's essence can once be extracted, there will be more than one way to reconstitute it.

Having teased apart various aspects of a text's essence, we can go on to consider the possible roles of future MT and AI in extracting these and conveying them to different languages and cultures. We'll consider several avenues for improvement in MT which promise to help in extracting these aspects of a text's essence – in Section 5.3, for enhancement of current, exclusively textually grounded MT (i.e., MT trained on text only), leading to extraction and delivery of high-quality *literal* translations; and in Section 5.4, in future *perceptually grounded* MT (i.e., MT trained on simulated perception, e.g., of audiovisual input, as well as text), which has the potential to extract associations, references, and other abstract aspects of a literary work to support *freer* translation.

With respect to perceptual grounding, I'll emphasize category formation and recognition, since I'll be claiming a crucial role for category recognition in extraction of abstract aspects of an artwork's essence. I'll discuss the learning of categories from perceptual instances as a key element in communication via linguistic symbols, and then extend the discussion to category-mediated translation. The latter, I'll suggest, can help MT move beyond literal translation toward greater freedom.

I'll remain cautiously optimistic regarding eventual development of automated esthetic judgment based on qualia, emotion, and consciousness, but will conclude that machine translation based on at least accurate semantic analysis of the original, increasingly augmented by perceptually grounded language processing supporting freer translation, is after all on its way. Artificial literary translation may after all support artistic and cultural preservation.

5.2 Multifaceted essences

We first reaffirm (1) that a source language text can have multiple aspects which jointly compose its essence, some easier to capture than others; and (2) that, once these have been extracted, there will be more than one way to reconstitute the original in a new language. As previewed, the infamous Nabokov-Wilson knockdown-dragout will cast a somewhat sanguinary light on these points.

The Eugene Onegin controversy

Nabokov's aim in translating *Eugene Onegin* was to enable non-Russian readers to appreciate Pushkin's masterpiece in all its depth: to comprehend every nuance of language, to command every fact whether historical or scientific, to follow every reference – in effect, to temporarily become the preternaturally knowledgeable Nabokov, knowing every relevant thing that he knew. Having taught Russian and European literature while excoriating the available translations as horribly inexact, he had concluded that the desired level of exegesis was incompatible with the attempt to create a rhyming and metered translation. Instead, he determined to offer "a pony" – a crib sheet or study aid à la Cliff's Notes, a painfully literal translation, accompanied by extensive annotations. In this case, "extensive" would be an understatement: whereas the translation itself would run to some 200 pages, the notes would fill four volumes! (This format – a poem encircled by a halo of commentary, suggesting embedded levels of consciousness and reality – would inspire his *sui generis* novel-cum-puzzle, *Pale Fire*.)

Edmund Wilson, a preeminent critic and ambitious author of the time, had helped Nabokov gain entry to the American literary world. Wilson was an omnivorous polymath, and in Nabokov's telling fancied himself as such, once having mystified party guests he mistook for lepidopterists by regaling them with butterfly arcana. He read in several languages including (apparently intermediate) Russian, and as his friendship with Nabokov grew, the two enjoyed discussing fine points of literature, for instance comparing – and amiably disagreeing about – English and Russian scansion.

But the close friendship, a rarity for the notoriously aloof Nabokov, ended abruptly following publication of the Pushkin translation. Wilson's review tore into Nabokov's treatment as fundamentally misguided: not only had his friend wasted his own undeniable gifts by eschewing an artistic rendering; he had failed on his own terms by abandoning any useful pedagogical literality in favor of recondite dictionary-only English terms beyond the ken of most students, where straightforward and easily understandable ones would have sufficed:

> ... the only characteristic Nabokov trait that one recognizes in this uneven and sometimes banal translation is the addiction to rare and unfamiliar words, which, in view of his declared intention to stick so close to the text that his version may be used as a trot, are entirely inappropriate here. It would be more to the point for the student to look up the Russian word than to have to have recourse to the OED for an English word he has never seen and which he will never have occasion to use. To inflict on the reader such words is not really to translate at all, for it is not to write idiomatic and recognizable English. ... He gives us, for example, *rememorating, producement, curvate, habitude, rummers, familistic, gloam, dit, shippon* and *scrab*. All these can be found in the OED, but they are all entirely dictionary words, usually labeled "dialect," "archaic," or "obsolete." Why is "Достойна старых обезьян" rendered as "worthy of old sapajous"? Обезьяна is the ordinary word for monkey.
>
> (Wilson, 1965)

Nabokov's exhaustive commentary, too, while undeniably conscientious, was mostly simply exhausting, Wilson sniped – pedantic to the point that a reader half expected Nabokov to provide genus and species specifics concerning the bear that appeared in Tatiana's dream. And – a parting shot intended as a *coup de grace* – Nabokov misinterpreted crucial story elements, most crucially in overestimating the protagonist's character.

Nabokov fired back and onlookers were wise to duck. "If told I am a bad poet, I smile; but if told I am a poor scholar, I reach for my heaviest dictionary." Wilson's critique was "… a polemicist's dream come true, and one must be a poor sportsman to disdain what it offers." First came a debunking of Wilson as an expert in the Russian language:

> A patient confidant of his long and hopeless infatuation with the Russian language and literature, I have invariably done my best to explain to him his monstrous mistakes of pronunciation, grammar, and interpretation. … Upon being challenged to read *Evgeniy Onegin* aloud, he started to perform with great gusto, garbling every second word, and turning Pushkin's iambic line into a kind of spastic anapest with a lot of jaw-twisting haws and rather endearing little barks that utterly jumbled the rhythm and soon had us both in stitches.
>
> (Nabokov, 1990, p. 248)

Nabokov then returned each Wilson shot, defending in turn *rememorating, producement, curvate, habitude, rummers, familistic, gloam, dit, shippon* and *scrab*:

> … once a writer chooses to youthen or resurrect a word, it lives again, sobs again, stumbles all over the cemetery in doublet and trunk hose, and will keep annoying stodgy gravediggers as long as that writer's book endures. In several instances, English archaisms have been used in my EO not merely to match Russian antiquated words but to revive a nuance of meaning present in the ordinary Russian term but lost in the English one. Such terms are not meant to be idiomatic. The phrases I decide upon aspire towards literality, not readability. They are steps in the ice, pitons in the sheer rock of fidelity. Some are mere signal words whose only purpose is to suggest or indicate that a certain pet term of Pushkin's has recurred at that point. Others have been chosen for their Gallic touch implicit in this or that Russian attempt to imitate a French turn of phrase.
>
> (Nabokov, op. cit., p. 252)

He dealt thus with *sapajou*, snapping shut a diabolically laid trap:

> In Mr. Wilson's collection of *bêtes noires* my favorite is "sapajou." He wonders why I render *dostoyno staryh obez'yan* as "worthy of old sapajous" and not as "worthy of old monkeys." True, *obez'yana* means any kind of monkey but it so happens that neither "monkey" nor "ape" is good enough in the context. "Sapajou" … has in French a colloquial sense of "ruffian," "lecher," "ridiculous chap." Now, in lines 1–2 and 9–11 of Four: VII ("the less we love a woman, the easier 'tis to be liked by her … but that grand game is worthy

of old sapajous of our forefathers' vaunted times") Pushkin echoes a moral-
istic passage in his own letter written in French ... The passage, well known
to readers of Pushkin, goes: "*Moins on aime une femme et plus on est sur de
l'avoir ... mais cette jouissance est digne d'un vieux sapajou du dix-huitième
siècle.*" Not only could I not resist the temptation of retranslating the *obez'yan*
of the canto into the Anglo-French "sapajous" of the letter, but I was also
looking forward to somebody's pouncing on that word and allowing me to
retaliate with that wonderfully satisfying reference. Mr. Wilson obliged – and
here it is.

(Nabokov, op. cit., pp. 255–6)

To recap Nabokov's justification in my own words, "I translated this everyday
Russian word for 'monkey' with a rare, gallic-flavored English word because the
original Russian passage purposely echoes one that Pushkin had previously written
in French, wherein he had used an uncommon French word for its humorous
associations."

And thus with the matter of interpretation:

My "most serious failure," according to Mr. Wilson, "is one of interpretation."
Had he read my commentary with more attention he would have seen that
I do not believe in any kind of "interpretation" so that his or my "interpretation" can be neither a failure nor a success. In other words, I do not believe
in the old-fashioned, naïve, and musty method of human-interest criticism
championed by Mr. Wilson that consists of removing the characters from an
author's imaginary world to the imaginary, but generally far less plausible,
world of the critic who then proceeds to examine these displaced characters as
if they were "real people." ... It is purely a question of architectonics – not of
personal interpretation. My facts are objective and irrefutable. I remain with
Pushkin in Pushkin's world. I am not concerned with Onegin's being gentle
or cruel, energetic or indolent, kind or unkind ... So much for my "most serious failure."

(Nabokov, op. cit., pp. 263–5)

Wilson got the worst of this first exchange of fire, to say the least.

Some years later, however, Douglas Hofstadter weighed in on Wilson's side.
Benevolent witness that he is, he recoiled at Nabokov's merciless fusillading of his
erstwhile friend (and of all authors of competing translations):

Amazingly, Nabokov's hardball savaging of his "old friend" Mr. Wilson
makes his criticism of [Pushkin translator Walter] Arndt look as wimpy as a
two-year-old's petulant toss of a Nerf ball. Could it be that Nabokov's took
as his model of "friendship" that between Onegin and Lensky, or (at least in
Pushkin's version) that between Salieri and Mozart, in which one "friend"
murders the other?

(Hofstadter 1997, p. 269)

Hofstadter was right to deplore Nabokov's self-certain, self-righteous nastiness. At the same time, empathy was due for the expat's defense of his honor and his subject's ground truth: Nabokov's utter mastery of Russian literature and his perceived duty, as the only major Russian writer in exile, to singlehandedly keep its flame burning were after all the very heartbeats of his identity, for which he would doubtless have fought any number of duels.[2]

However, Hofstadter's main point was that Nabokov's renunciation of the quest for a translation both accurate and artistic was premature, not to say faint-hearted. Attempting an existence proof that such a translation could after all be created, Hofstadter presented for comparison (Hofstadter, 1997, pp. 242–5) four rhymed and metered translations – by Oliver Elton (revised by A. D. P. Briggs) (Pushkin, 2016); Charles Johnston (Pushkin, 1979); James Falen (Pushkin, 1995); and Walter Arndt (Pushkin, 2002) – inviting readers to judge the results for themselves (while referring to Nabokov's literal version to assess accuracy). You'll find several of these translation samples in Appendix I.

For me, Falen is the winner, in both accuracy and apparently effortless esthetics. Hofstadter agreed, though he subsequently undertook his own translation of *Eugene Onegin*, also sampled in Appendix I.

But again, our goal at present – beyond guilty enjoyment of the reality-TV duel[3] – is to enquire whether any of these renderings, or any likely to follow, would convince Nabokov's ghost that a rhyming and metrical translation could be worthy of Pushkin's artistry while so precise in its meaning that it could serve as the reference point for Nabokov's exhaustive annotation. The answer – I feel confident in channeling Nabokov's shade – is no. His perfectionism, his drive to nail the literal meanings, thus providing the stoutest possible hook on which to hang the full treasure of his knowledge, was simply too exacting.

So how are we to conclude? Were Wilson and Hofstadter right to hold out for a translation at once quite close to the literal meaning and artistically satisfying? Or was Nabokov right to maintain that maximum literal accuracy is incompatible with maximum artistry? The judges have reached their decision, and the answer is …

I'm reminded of the classic joke: A rabbi adjudicates between two congregants. The first gives his side of the story. The rabbi rubs his chin and says, "You're right." The second gives his. The rabbi raises his eyebrows and says, "*You're* right." The rabbi's wife exclaims, "But they can't both be right!" To which the rabbi nods sagely and responds, "And you're right, too."

Both Nabokov and Wilson were right. As forecast, I believe that the *Onegin* controversy and the various translation versions it has prompted do reconfirm, when taken together, that the essence of this translation (and by extension, most others) is not unitary but many-faceted: it has several factors that can be teased apart, viz. literal meaning, references, associations, knowledge of the context, rhyme, meter, and more.

From this viewpoint, Nabokov was right that a trade-off must be recognized: absolute maximization of one or more of these factors usually does preclude absolute maximalization of others. Some compromise will be necessary if the attempt is made to have it all. In other words, *a literary translation, or any other multi-factored translation, should be viewed as an optimization problem, in which each factor's score will usually be imperfect but the best* combined *score can be sought.*

On the other hand, Wilson and Hofstadter were right that *the determined attempt to optimize a combined score can yield overall results better than expected* (and better than Nabokov was prepared to admit).

Le ton beau de Marot

Additional examples from Hofstadter show that one or more aspects of a text's essence can be held steady (fixed or "clamped") during translation while varying others, even to the point of abandoning or replacing them. Such permutations will yield a series of variant translations. In this case, the literal meaning of a short and sweet French poem (Clément Marot's "*A une Damoyselle malade*") was retained while ringing changes on the remaining aspects – references, associations, register, rhyme scheme, meter – to obtain quite a few stylistically varying translations. Several appear in Appendix II.

We're seeing that several translations can be offered side by side to communicate different aspects of the source text. This insight seems in retrospect like a rather obvious cutlass with which to cut the gordian knot: copious ink – and the heart's blood of a beautiful friendship – might have remained unspilled if some Solomon had been on the spot to recommend it to Wilson and Nabokov.

I put in my own mustard. Having avidly read Hofstadter's take on the *Onegin* confrontation, I showed up at a talk he gave at UC Berkeley sometime in the eighties, excited to present a gift copy of *Mangajin*, a magazine for learners of Japanese. In it, numerous *manga* (comics) appeared each month with translations of all captions in the following unique format: (1) original Japanese; (2) romanized transcription; (3) term glosses; (4) literal translation; and (5) freer/more natural translation.

> うむ、この 舌ざわり この 歯ごたえ、比較 に ならないうまさだ！！
> *Umu　kono shitazawari kono hagotae　hikaku ni naranai umasa da*
> Uh-huh this tongue-touch this tooth-response no-comparison tastiness is
> "Yes, this texture, this firmness, it's a tastiness that's no comparison."
> **"Yes, this texture, this firmness, it's so tasty there's no comparison!"**

Eugene Onegin, I hurriedly suggested, could certainly be treated likewise. Then I rushed off to an errand I could have rescheduled, thereby missing an intimate dinner with Hofstadter planned by the Linguistics Department – one of the real regrets of my life.

5.3 Textually grounded MT

The juicy *Eugene Onegin* controversy has supported our reaffirmation that the essence of an artistic text is not monolithic and indissoluble. Rather, we can recognize multiple aspects of that essence (or, if you like, multiple essences): the text's literal meaning; its structural aspects like rhyme and meter; the knowledge and emotion behind it; and a wide variety of its stylistic, associated, and connotational elements. But can these aspects be captured and transferred via MT and AI?

Among the several distinguishable aspects of a text's essence, one stands out as first among equals: the literal meaning. A putative translation may be stylish and rich in associations, and it may rhyme and scan wonderfully, but if it doesn't at least convey the intended meaning, or at least one closely related, its claim to be a translation at all is doubtful. So, in the spirit of first things first, we'll begin with literality: can we enable a strictly literal machine translation exhibiting accuracy high enough for even literary applications? I think that this will be possible; and I think it can be done using only textually grounded techniques. Accordingly, this section will remain with these methods, postponing discussion of perceptually grounded MT until Section 5.4.

Two clarifications at the outset: first, hasn't machine translation already achieved Fully Automatic High Quality (FAHQ) – literal translation of sufficient quality for delivery to the end user without human intervention for checking and correction? Unfortunately, not yet: while recent progress has been dramatic, claims of near-human yet human-free performance should be liberally salted prior to consumption. Especially where complex texts are concerned, many errors – numerous nits and not a few howlers – are still to be expected. Second, isn't neural MT already "perceptually grounded"? After all, neural networks suggest artificial brains, and aren't brains in the business of perceiving? But again, no: by perceptually grounded, we mean *trained upon* visual, auditory, or other sensor input in addition to or instead of text. Most current research in neural machine translation (NMT) still employs only text during training, and all commercial systems still follow suit. (But what about speech translation? Yes, those systems do receive audio input intended to train for recognition of speech; but our topic here is rather the translation of literary texts, and we need to distinguish current speech translation methods, which include in their training processes only sensory input designed to prompt formation of phonological concepts, from future text translation methods which will aim to train concepts of all sorts.)

We can move on, then, to discuss directions for improvement grounded only in text. We'll group them under two main headings: further development of *textually grounded neural* methods per se; and integration of various knowledge sources.

5.3.1 Directions for improvement

We'll survey two paths for improvement of neural (but not perceptually grounded) MT techniques: making big data bigger, with a vital role for humans in the loop; and increasing integration of semantic elements into MT, including symbolic (handmade), vector-based, and neural-network-internal semantics.

5.3.1.1 Big data and interactive MT

Big data has undeniably played a decisive role in improvement and scaling of machine translation to date. More data is better data, and plenty more is on the way (though the rise of neural technology is increasing the importance of data quality as well as quantity). Text translation data – both parallel and monolingual – is already massive but will be increased through a virtuous circle: as systems improve, they will be used more, thus producing still more data, and so on.

The new translation data must be correct to be most useful for further machine learning. Correction has traditionally been made by expert post-editors, but the current trend is toward end user correction of preliminary translation results. Google Translate has made a strong beginning in this direction by enabling users to suggest improvement of preliminary machine-made translations via the company's Translate Community. And just as more data is desirable, so are more corrections. At present, however, there is a barrier which limits the crowdsourcing community: to correct translations most effectively, users must have at least some knowledge of both the source and target language – a handicap especially for less-known languages. However, verification and correction techniques designed for monolinguals could greatly enlarge the feedback base.[4]

This exploitation of user-generated feedback jibes with three general trends: (1) toward *interactive* translation methods; (2); toward *continual* learning, as opposed to batch training based on static corpora; and (3) toward attempts to pry open computational black boxes.

Interactivity. *The interactive trend is of special interest for future literary translation*: humans can intervene, so that the system moves toward machine-aided human translation, or human-aided machine translation: rather than wait to post-edit the finished but rough results of machine translation, translators can intervene during the translation process (or vice versa).[5]

Dynamic updating. MT systems can improve very rapidly (and can be quickly customized for specific use cases) if translation models can be updated after each correction, and even more so if corrections are made interactively rather than after translation of an extensive text.[6]

Opening black boxes. The AI field is seeing an increasing effort to maintain traceability, comprehensibility, and a degree of control – keeping humans in the loop to exert that control and avoid errors. In Google's experiments with driver-less cars, for instance, designers have struggled to maintain a balance between full autonomy on the car's part, on one hand, and enablement of driver intervention on the other. Rather than treat MT and other AI systems as black boxes – as oracles whose only requirement is to give the right answers, however incomprehensibly they may do it – the aim is instead to build windows into artificial cognitive systems, so that we can follow and interrogate, and to some degree control, internal processes. Granted, the black box path is tempting: it is the path of least resistance, and in any case organic cognitive systems have until now always been opaque – so much so that behaviorism ruled psychology for several decades on the strength of the argument that the innards were bound to remain opaque, so that analysis of input and output was the only scientifically respectable way forward. However, because artificial cognitive systems will be human creations, there is an unprecedented opportunity to peer within them and steer them. As we build fantastic machines to deconstruct the Tower of Babel, it would seem healthy to remember the Sorcerer's Apprentice: best to have our minions report back from time to time, and to provide them with a HALT button.

5.3.1.2 More semantics

(Seligman, 2019) discusses the evolving treatment of semantics in MT, with attention to symbolic (handmade) semantics and ontologies; vector-based semantic approaches; and semantics internal to neural networks. I'll briefly reprise that discussion here, since increasing exploitation of semantics will be especially relevant to extraction of meaning, whether more or less literal, in future automatic literary translation. Keep in mind, however, that all three approaches currently remain textually grounded. We're postponing discussion of perceptually grounded semantics until Section 5.4.

Symbolic semantics. Explicit symbolic representation has undergone a marked decline since the rule-based era of MT. However, a comeback seems likely, at least for some purposes.

In the first place, better results may sometimes be obtained: by reference to previously translated corpora, translation of Japanese *kyouto no kaigi* or *toukyou no kaigi* could be enhanced – yielding *conference in Tokyo* or *Tokyo conference* rather than a stilted direct translation like *conference of Tokyo* – through exploitation of symbols drawn from ontologies (collections of semantic or conceptual symbols), indicating that *kyouto* and *toukyou* are examples of the CITIES class and that *kaigi* is an example (or subclass) of MEETINGS.[7] Other advantages relate to universality and interoperability. Regarding universality, the original argument for interlingua-based MT in the rule-based era was after all that the number of translation paths could be drastically – in fact, exponentially – reduced if a common pivot could be used for many languages. In that case, the meaning representation for English would be the same as that for Japanese or Swahili, and all analysis or generation programs could be designed to arrive at, or depart from, that same pivot point. And concerning interoperability, the same representation could be shared not only by many languages but by many machine translation systems. Thus the ambition to overcome the Tower of Babel among human languages would be mirrored by the effort to overcome the current Babel of translation systems.

A common meaning representation, beyond bridging languages and MT systems, could also bridge natural language processing tasks. And in fact, we do see explicit semantic representation taking hold now in tasks other than translation. Google, for example, has already begun to make extensive use of its Knowledge Graph ontology[8] in the service of *search*. "Thomas Jefferson" is now treated not only as a character string, but as a node in a taxonomy representing, e.g., an instance of the PERSONS class, and of its LEADERS subclass, and of its

PRESIDENTS sub-subclass, and so on. This knowledge guides the search and enables more informative responses. Similarly, IBM's Watson system uses its own Knowledge Graph, this time in the service of *question answering* – initially focused especially upon the healthcare domain.[9] Unsurprisingly, Google and IBM presently use their own ontologies. However, eventual movement toward a common standard seems likely: one semantic representation that could bridge languages, tasks, and competing or cooperating organizations. Meanwhile, efforts to inter-map or mediate among competing taxonomies also seem likely.

For automatic or computer-aided literary translation, one would devoutly wish to model the encyclopedic literary and cultural knowledge of a Nabokov. And in fact, inroads have already been made: Watson has provided an existence proof by beating the world champions in Jeopardy, a game requiring far-reaching knowledge and sophisticated processing of language encompassing associations, puns, and so on. To enable automatic play, systems internalized entire encyclopedias and other knowledge sources, albeit in textually grounded form.

Vector-based semantics. During the reign of statistical machine translation (SMT) now ending, symbolic semantic representation was utilized only rarely. In compensation, *vector-based* semantic treatments gradually became influential. These aim to leverage the statistical relationships among text segments (words, phrases, etc.) to place the segments in an abstract space, within which closeness represents similarity of meaning (Turney and Pantel, 2010). Intuitively, words that occur in similar contexts and participate in similar relations with other words should turn out to be semantically similar. The intuition goes back to Firth's declaration (1957) that "You shall know a word by the company it keeps," and has been formalized as the *distributional hypothesis*. The clustering in this similar-neighbors space yields a hierarchy of similarity relations, comparable to that of a hand-written ontology. Historically, the vector-based approach grew out of document classification techniques, whereby a document can be categorized according to the words in it and their frequency. The converse was then proposed: a word or other linguistic unit can be categorized according to the documents it appears in, or more generally, according to surrounding or nearby text segments of any size – minimally, just the few words surrounding it. (Turney and Pantel, 2010, survey the various sorts of contexts explored to date.) See Alkhouli, Guta, and Ney, 2014, for an example of experimental use of vector-based semantics to improve an SMT system.

Vector-based semantic approaches will probably continue to be used in combination with symbolic and semantic-network-based methods to contribute to overall MT quality, and can in this way help move the needle toward sufficient accuracy for literary projects.

Neural semantics. Neural networks were born to learn abstractions. The "hidden" layers in a neural network, those that mediate between the input and output layers, are designed to gradually form abstractions at multiple levels by determining which combinations of input elements, and which combinations of combinations, are most significant in determining the appropriate output. We can view each abstraction level as a stage in a chain of implied "rules." Rules close to the input, at the bottom of the neural network, use surface elements specific to particular inputs as their "premises," while those at higher layers use "premise" combinations taken from many inputs. The more hidden layers, the more levels of abstraction become possible; and this is why *deep* neural networks are better at abstracting than shallow ones. This advantage has been evident in theory for some time; but deep networks only became practical when computational processing capacity became sufficient to handle multiple hidden layers.

Where machine translation is concerned, this hidden learning raises the possibility of training neural translators to develop internal semantic representations automatically and implicitly (Woszczyna et al., 1998). A new neural-network-based approach to semantics then suggests itself: *within a network, nodes or pathways shared by input elements having the same translation or translations can be seen as representing the shared meanings.* Input elements sharing a translation can originate in a single source language (when in that language the source elements are synonyms in the current context) or in several source languages (when across the input languages in question the source elements are synonymous in their respective contexts). And in fact, the shared translations, too, can be unilingual or multilingual.

Thus, if translation is trained over several languages, semantic representations may emerge that are abstracted away from – that become relatively independent of – the languages used in training. Taken together, they would compose a *neurally learned* interlingua, a language-neutral semantic representation comparable to the *handmade* symbolic interlingua developed over many decades (Uchida, 1986).[10] A successful neural interlingua could facilitate handling of underresourced languages, thus opening a path to truly universal translation at manageable development costs. Several teams have begun work in this direction (Le et al., 2016)[11,12], and early results are already emerging: Google, for instance, has published on "zero-shot" NMT, so named because the approach allows translation between languages for which zero bilingual data was included in training corpora (Johnson et al., 2016); and SYSTRAN, in a similar spirit, has already announced combined translation systems for romance languages.[13] Zero-shot NMT works because the encoding (analysis) phase of translation has been generalized across all currently trained source languages, while the decoding (generation) phrase has similarly been generalized across all currently trained target languages. Thus any current source can be paired with any current target. Expectations would be low, however, if completely untrained source or target languages were tried.

5.3.2 Knowledge source integration

The history of spoken language translation, recounted in (Seligman and Waibel, 2019), is an interesting chapter of the wider MT story. In the course of it, several large projects attempted to integrate multiple knowledge sources such as discourse analysis, prosody, topic tracking, and so on. The results were mixed, since the infrastructure – computational, networking, architectural, and theoretical resources – necessary for such complex integration did not yet exist. These resources do exist now, though, as witness the progress of IBM's Watson system,[14] already mentioned. Watson calls upon dozens of specialized programs to perform question-answering tasks, relying upon machine learning techniques for selecting the right expert for the job at hand. In view of these advances, the time seems right for renewed attempts at massive knowledge source integration in the service of machine translation. Some sources may be programmed as neural networks, while others may retain standard or statistical programming.

5.3.2.1 Discourse analysis

For general translation purposes, knowledge sources relating to discourse structure should prove useful. For instance, to enhance treatment of coherence beyond the clause level, knowledge of discourse relations (like CAUSE, HOWEVER, and FOR EXAMPLE) could be brought in. (Seligman, 1991, studies such relations in the context of natural language generation.) For handling of pragmatic aspects of texts, speech act analysis programs could be used. To tighten processing of specific topics, topic tracking programs could participate. All these are discussed (in the context of speech translation) in Seligman, 2000.

5.3.2.2 Combinatorics

I've discussed the desirability of human involvement in the automatic translation process. This interactivity implies selection among alternative translations or fragments suggested by the MT program. For literary translation, the need to consider many alternatives becomes particularly acute. In composing a rhymed and metered version of a poem or song, for example, the human translator may try out dozens of versions per line while seeking those that rhyme and scan; but of course, prose, too, calls for myriad esthetic choices. Given that thousands of paraphrases can be generated for even an everyday sentence of medium length (Mel'chuk and Zholkovski, 1970, provide a classic demonstration), vast possibility spaces will sprout which human captains can navigate only via carefully designed interfaces. (Seligman, 1991, outlines a multiple-worlds mechanism for tracking branching possibilities during natural language generation, comparable to the data structures used to track board positions in automatic chess players.)

I mentioned rhyme and rhythm. Their inclusion in a translation system will call for dedicated resources (pending the arrival of very general AI). These could make use of existing rhyming dictionaries and metric analyzers – or create customized new ones – to filter out and highlight (or actively seek, by pruning search paths) translation possibilities which rhyme with specified lines and/or which yield the specified meter for the current line.

And of course, many additional specialized modules could be designed to apply various esthetic constraints: brevity checkers to model the soul of wit (without yet having any of either); repetition eliminators; filters for register, domain, or style; and so on. (For now, we're postponing discussion of true esthetic judgment in favor of such programs exploiting rules of thumb, whether handwritten or acquired through machine learning.)

5.3.3 Prospects

The upshot: *Textually grounded MT development efforts already underway, when integrated with various modules already existing or feasible, are likely to enable literal translations of sufficient quality for literary use cases* – providing they are applied in a well-designed architecture enabling human oversight for interactive correction of, and selection among, translation candidates. With a fair wind, a hyper-literal human-aided machine translation of *Onegin* like Nabokov's, complete with exhaustive annotation, should be possible.

It's unlikely, however, that sufficient accuracy or flexibility for relatively free translation of literature can be achieved by textually grounded systems *without* human participation. The programs may simulate knowledge and comprehension of the world to a greater or lesser degree, but without perceptually grounded worldly experience will remain limited in their capacity to suggest metaphors, make associations, or exhibit other marks of human-like free translation. The freer a translation attempt, the less the likelihood of success at this textually grounded level; so a hyper-free *Onegin* translation like Hofstadter's is not expected if also *human*-free. Instead, the hope would be to use an accurate literal translation plus annotations as jumping off points for various free human translations.

Let's progress, therefore, to consideration of perceptually grounded translation, where freer translation – extraction of essence aspects beyond that of literal meaning – becomes more likely.

5.4 Perceptually grounded MT

Again, the semantic representations discussed earlier – explicit symbolic, vector-based, and early neural-network-based – have in common that they have until now been based (that is, trained) on text alone. In contrast, we now turn to the possibility of *perceptually grounded semantics* – semantics based upon actual (even if computationally simulated) experience, whether visual, auditory, tactile, or other. Our brief will be that this sort of meaning entails the formation and recognition of categories or classes at many levels of abstraction; that multi-level class recognition will be crucial for extracting the essence of a situation; and that this recognition in turn will be, well, essential for achieving the fluid associations and abstractions necessary for progress beyond purely literal translation.

In making this argument, we begin by considering the role of category formation in a prototypical scenario of monolingual symbolic communication. We can then extend the example toward cross-lingual – that is, translated – communication. The penultimate step will be to outline our approach to neural (or simulated neural) category formation. In the final step, we'll wave our hands toward exploitation of neural categories for the fluid associations characteristic of relatively free translation.

5.4.1 Perceptually grounded symbolic communication

We can imagine a computational system that learns from visual, audio, or other sensor-based examples to recognize members of the category CATS, thereby internalizing this category;[15] learns from examples to recognize members of the graphic category NEKO-KANJIS (the Japanese character 猫, symbolizing the meaning "cat"), thereby internalizing this second category; and learns from examples to associate the two categories in both directions, so that activation of CATS triggers activation of NEKO-KANJIS and vice versa. We can also imagine a second computational system with similar learning mechanisms that learns likewise, but based on completely different examples. And finally, assuming that at least one of the systems can learn to generate and transmit *new* instances of NEKO-KANJIS, we can imagine communication between the two systems mediated by transmission of such instances and confirmed through some objective functional test, such as reliable selection from a barnyard lineup. The argument then is that, to both systems, instances of 猫 have a kind of meaning absent from handmade, vector-based, or even neural-network-based "semantic" constructs divorced from (even artificial) perception.

This linguistic communication scenario could in fact be implemented using current technology. The DeepMind neural net technology acquired by Google can indeed form the category CATS (minus the label) based upon perceptual instances in videos (much as the perceptual systems of self-driving cars are daily internalizing and refining categories like PERSONS, VEHICLES, etc.). And as for the learning of communicative symbols like NEKO-KANJIS, in fact every speech recognition or handwriting recognition program already forms implicit categories such that a new instance is recognized as belonging to the relevant category. What remains is to learn the association between categories like CATS and NEKO-KANJIS, and then to demonstrate communication via the symbol categories between computers whose respective learning has depended upon unrelated instances.

5.4.2 Perceptually grounded translation

We can extend this story of perceptually grounded semantics to translation by assuming that, while one computational system learns an association with NEKO-KANJIS, the other learns a different linguistic symbol category instead, say that of the written word "cat" in English, call it CAT-GRAPHEMES. Then for communication to take place, a kanji instance must be replaced during transmission by an instance of that graphic class (or vice versa). *If the replacement involved activation in a third system – the translator system – of a perceptually learned CATS class associated with both the learned source and target language symbols, then the translation process as well the transmission and reception would be perceptually grounded.*

5.4.3 Category learning as intersection of percepts

Our story about perceptually grounded communication via linguistic symbols depends upon the ability to learn categories (classes) through experience of multiple instances – for instance, to learn the class of CAT.PERCEPTS by experien-cing CAT.PERCEPT.1, -2, -.3, and so on.

We've assumed that brains, on perceiving multiple instances, carry out the necessary generalization or abstraction for class formation *somehow*, postponing consideration of the where and how. Likewise, association between general and symbolic classes has been assumed to occur through temporal proximity, but no specific mechanisms have yet been suggested. To support upcoming discussion concerning the role of categorization in free translation, I'll sketch an approach to neural category formation now and hope to fill it out in forthcoming publications.

The guiding hypothesis will be that category formation occurs when the brain discovers the intersection of multiple instances. The instances, we'll assume, are represented both globally and locally in the brain – globally because percep-tual information travels via neural pathways from various locations in the brain specialized for vision, hearing, and so on; and locally because that distributed information arrives at, and is integrated within, narrowly localized sections of associative areas in the brain. The hypothesized high points are as follows:

- Memory of an instance (for example, that of a specific cat at a specific time) is formed at such an integration point (for instance, binding together the shape, sound, color, and size of the percept). We can picture the instance memory as the nexus or convergence point of several neural pathways coming from the perceptual areas (Figure 5.1).

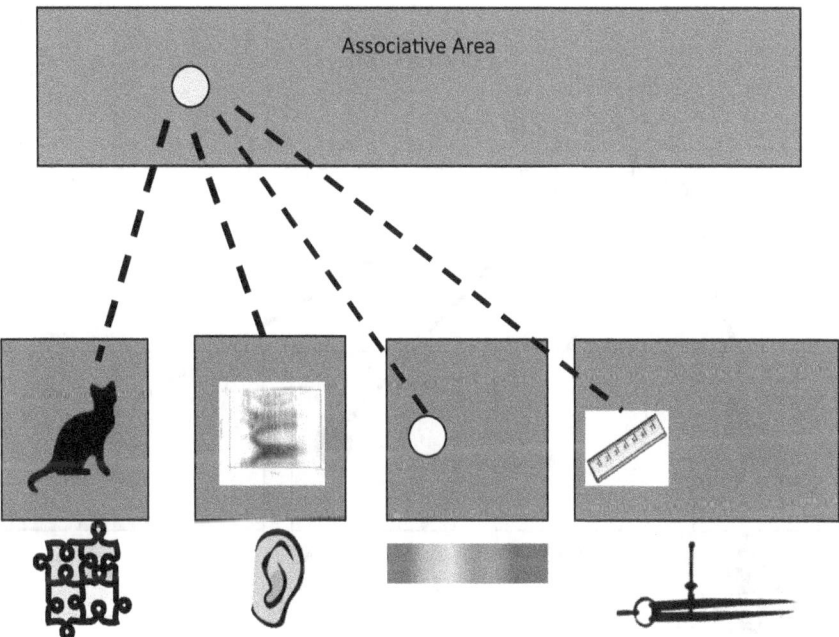

Figure 5.1 A perceptual instance (small white cat), showing neural value areas, neural pathways, and a neural associative area.

- Once formed, the memory can be activated by perception of a new instance sharing sufficient perceptual input features (for example, having similar shape and sound). Such activation of the original memory can in turn activate all of the originally associated features (such as the original color and size). *NOTE: This is the account of class/concept recognition which will be central to our speculations concerning neural abstraction.*
- Some features of the new instance may differ from those of the original instance. (For example, the original may have been a small white cat, whereas the new one is a big black one.) The difference may be sufficient to prompt formation of a separate memory of the new instance (Figure 5.2).
- Whether or not a new instance memory is formed, the shared features of the original instance and the new one (for instance, their respective shapes and sounds) are reinforced, since the relevant neural pathways have been deepened or strengthened through repeated use. And – we hope you guessed – *the paths more worn, because they're more traveled, now constitute our newborn neural class or category* (Figure 5.3).
- The several instances contributing to a category all have their convergence points in the same or nearby associative area. Consequently, the resulting category, too, has a localized nexus in the same associative area, perhaps arising from a merger or overlap of those of the several instances.
- Categories representing linguistic symbols, for example, NEKO-KANJI. PERCEPTS will be learned through intersection of multiple instances in the

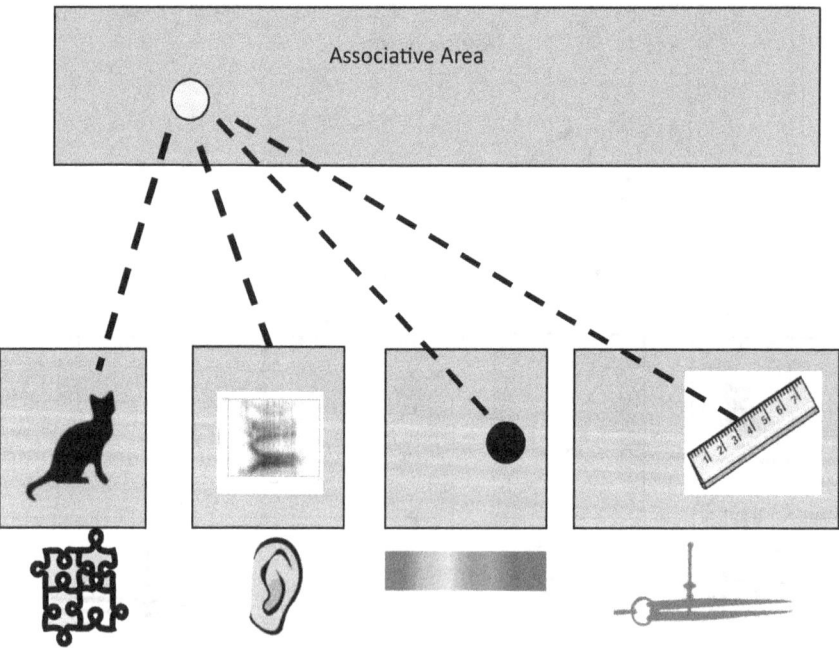

Figure 5.2 A second perceptual instance (big black cat), again showing neural value areas, neural pathways, and a neural associative area.

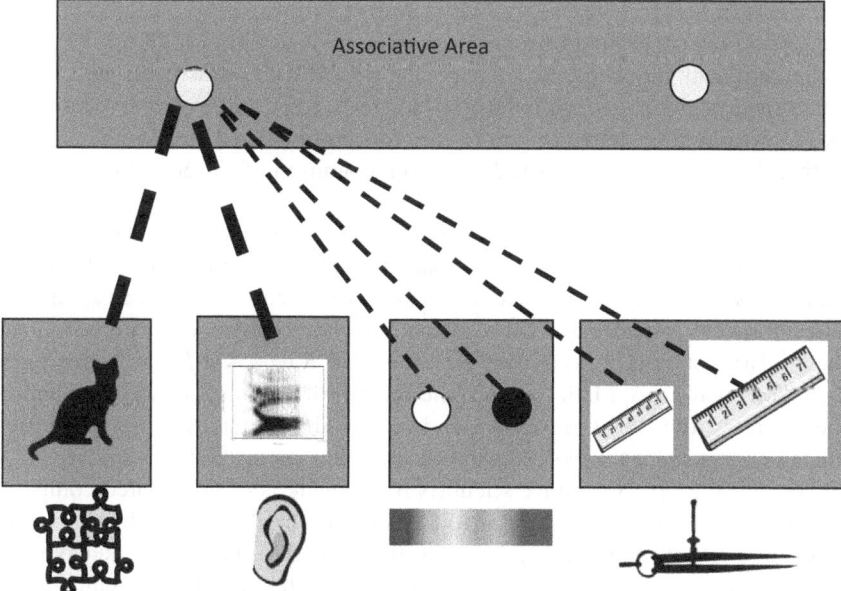

Figure 5.3 The new CAT.PERCEPTS category, formed through reinforcement of common elements of two instances.

same way. (as suggested by the right-hand circle in the associative area of Figure 5.3)

- Existing classes can play parts in the learning of additional classes, so that classes can be formed at many levels of abstraction, and involving various sensory modalities.
- Since the neural categories representing, e.g., CAT.PERCEPTS and NEKO-KANJI. PERCEPTS are localized, association between them (likely based on temporal proximity of their respective activations) is facilitated: the association can be at the category level (as might be depicted with a dashed line linking the circles in the associative area of Figure 5.3), and thus need not interlink all of the respective features in a potentially explosive peer-to-peer manner.

Armed with this lightning sketch of class formation and association in brains (natural or artificial), we return to this chapter's central theme: What are the essences of a text, and how can they be extracted by automatic translators?

5.4.4 Essence extraction revisited

In Section 5.3, we suggested that high quality *literal* translations, accompanied by extensive notes, could be provided by improved but still textually grounded MT, with assistance from even monolingual humans. We're now anticipating growing capacity for *freer* translation based on perceptual grounding. Essence extraction in brains, I believe, can be identified with category recognition; and I've hypothesized, in broad strokes, how that learning occurs and how it can be computationally simulated. As a first result, we can see how associations among categories can give rise to symbolization, and thus, when automatized, to human-like inter-computer communication and concept-mediated translation. Going further, however, we open a window onto recognition of all sorts of essences at all levels of abstraction.

And here Douglas Hofstadter will once again come to our aid. Throughout *Surfaces and Essences* (Hofstadter and Sander, 2013), he and Emmanuel Sander explore essence recognition as a crucial cognitive process – at root, an ability to find similarities and analogies among situations. For example, one can recognize the similarity of plots between *West Side Story* and *Romeo and Juliet*. In this analogy, Tony is Romeo; Maria is Juliet; the Jets are the Montagues and the Sharks the Capulets; the Bronx is Verona; and so on. Hofstadter contends that little human cogitation could proceed without the ability to analogize in this way.

He isn't the first. Cognitive scientists have for decades highlighted comparable analogy recognition as central to thinking, inspired especially by the work of George Lakoff and Mark Johnson (2003) on the ubiquity of metaphors. Computational studies have followed, beginning with attempts, for example, by (Jacobs, 1985) to intermap the elements of analogous structures in handmade semantic networks. Hofstadter has carried the ball further (to use a sports metaphor), attempting to model the making of very fluid, human-like analogies, but still in a handmade programming style.

My suggestion is to generalize the main cognitive operation in question. To repeat, I think we can view analogizing as a special case of categorization or classification – that is, of recognition and exploitation of similarities or overlaps among two or more instances. In traditional analogies and metaphors, the instances to be categorized are relatively complex structures or situations entailing non-trivial intermapping of elements – as in the earlier-mentioned similarity of star-crossed romantic plots. But we can see categorization of simpler structures as fundamentally similar to that of more complex ones in their formation and use.

And I think that category recognition is precisely the essence extraction process that we've been pursuing – the reminding *process, the one which activates a class in response to presentation of a new, similar-but-not-identical instance*; the one which zeroes in so impressively, even uncannily, on the essence of a concept.

From the neural network perspective, and from the complementary perspective of our category learning cartoon, the process of extracting many sorts of essences becomes more understandable. It may well exploit the multiple levels of abstraction inherent in deep (multi-layer) neural networks. In my suggested neural view, category recognition leverages the potential of pathways to converge at many locations, representing many sorts of similarities; and similarities can be represented recursively by establishing categories of categories.

Perceptually grounded natural language processing will, I believe, eventually enable automatic translation to simulate translation beyond the literal, as it will add manifold associations, analogies, similes, metaphors, and connotations to the mix – all manner of abstractions.

5.5 Conclusion

We began with the image of translation as extraction of essence and, as spectators of a notorious literary duel, confirmed that a literary text will generally yield many essences, not just one. However, a standout among these several essences is the text's literal meaning, as championed (under intense fire, but giving as good as he got) by Vladimir Nabokov – who also insisted that the core meaning is the nexus of a halo of facts, references, associations, etc. We suggested that further development of MT along current, textually grounded lines could, with even monolingual human assistance, provide accurate literal translations plus Nabokovian commentary. Humans should be able to depart from these to augment, elaborate, and vary, thus producing freer translations. Finally, we looked ahead toward perceptually grounded MT, which should eventually enable automatic production of freer translations, amenable to human evaluation, selection, and curation, by adding abstract elements – associations, analogies, similes, metaphors, connotations, and more. The process of essence extraction, I argued, is in neural terms the all-important process of categorization – of the learning and recognition of classes; and we can begin to see how this process operates in brains and how it can be computationally simulated.

In the last analysis, however, fully automatic production of truly high-quality translations will require mature esthetic judgment. When examining millions of possible versions at superhuman speed, some selection criteria beyond rules of thumb will be needed. Selection must ultimately depend not only on breadth of literacy – computers can read much more than you can, potentially speed-reading the Library of Congress in a sitting – but on feeling, on qualia, on emotion, on consciousness or something like it.

Is such judgment even possible in computers? I think so, in principle. Brains and bodies are material things, after all, so there should be no principled obstacle to doing everything they do in other media. But we're far from understanding how to automate feeling – and from evaluating how we should. Pending such understanding and such evaluation, human co-translators will have to supply the emotion and esthetics.

Sustenance of cultures will depend in part upon transmission of a culture's linguistic artifacts – of its literature. To help preserve and transmit a work of literary art, our automatic minions will travel down its secret stem, reach the root, and feed upon it. The nutrients they return to us will increase with time.

Notes

1 Also, the "dead" and "alive" images are more apt for document translation than for real-time translation, such as translation of spoken conversations; but real-time exchanges – my R&D specialty – also have a role in language preservation.

2　His father did in fact call out the editor of a powerful Rightist newspaper for commissioning a "scurrilous piece" containing unpardonable insinuations; but the offender apologized and called the duel off – upon which his father's "large cool hand resting on my head did not quaver," as Nabokov proudly recalled in his memoir *Speak Memory* (Nabokov, 1999, page 150).

3　The showdown is entertaining enough to have spawned an entire book (Beam, 2016).

4　Research on verification and correction by monolinguals has since 2002 been the main thrust of my own R&D company, Spoken Translation, Inc.; and exciting recent work on interactive *neural* translation has been reported by, for example, Peris et al. (2016).

5　As recently promoted, for instance, by the machine translation startup Lilt.

6　See, e.g., www.statmt.org/survey/Topic/IncrementalUpdating.

7　See https://en.wikipedia.org/wiki/Example-based_machine_translation.

8　See https://en.wikipedia.org/wiki/Knowledge_Graph.

9　See http://researcher.ibm.com/researcher/view.php?person=us-anshu.n.jain.

10　See also, e.g., www.undlfoundation.org/undlfoundation/images/cfp%20-%20unl%20programme.pdf.

11　See, e.g., www.slideshare.net/eraser/googles-multilingual-neural-machine-translation-system-enabling-zeroshot-translationandwww.kurzweilai.net/googles-new-multilingual-neural-machine-translation-system-can-translate-between-language-pairs-even-though-it-has-never-been-taught-to-do-so.

12　See also www.aclweb.org/anthology/N16-1101.

13　See http://forum.opennmt.net/t/training-romance-multi-way-model/86.

14　See www.ibm.com/watson/.

15　Pun intended.

References

Alkhouli, Tamer, Andreas Guta, and Hermann Ney. 2014. "Vector space models for phrase-based machine translation." In *Proceedings of SSST-8, Eighth Workshop on Syntax, Semantics and Structure in Statistical Translation*. Doha, Qatar, October 25, pp. 1–10.

Beam, Alex. 2016. *The Feud: Vladimir Nabokov, Edmund Wilson, and the End of a Beautiful Friendship*. New York: Pantheon.

Firth, John Rupert. 1957. "A synopsis of linguistic theory 1930–1955." In *Studies in Linguistic Analysis*. Oxford: Philological Society 1 (32). Reprinted in F.R. Palmer (Ed.), 1968, *Selected Papers of J.R. Firth 1952–1959*. London: Longman.

Hofstadter, Douglas. 1997. *Le Ton Beau de Marot: In Praise of the Music of Language*. New York: Basic Books.

Hofstadter, Douglas and Emmanuel Sander. 2013. *Surfaces and Essences: Analogy as the Fuel and Fire of Thinking*. New York: Basic Books.

Jacobs, Paul. 1985. A Knowledge-Based Approach to Language Production. Ph.D. thesis. University of California, Berkeley, Computer Science Division Technical Report UCB/CSD 86/254.

Johnson, Melvin, Mike Schuster, Quoc V. Le, Maxim Krikun, Yonghui Wu, Zhifeng Chen, Nikhil Thorat, Fernanda Viégas, Martin Wattenberg, Greg Corrado, Macduff Hughes, and Jeffrey Dean. 2016. "Google's Multilingual Neural Machine Translation System: Enabling Zero-Shot Translation." https://arxiv.org/abs/1611.04558.

Lakoff, George and Mark Johnson. 2003. *Metaphors We Live By*. Chicago: University of Chicago Press.

Le, Than-He, Jan Niehues, and Alex Waibel. 2016. "Toward multilingual neural machine translation with universal encoder and decoder." In *Proceedings of the International Workshop on Spoken Language Translation (IWSLT) 2016*. Seattle, WA, December 8 9.

Mel'chuk, Igor A. and A. K. Zholkovski. 1970. "Toward a functioning meaning—Text model of language." *Linguistics*, 57 (1970), 10–47.

Nabokov, Vladimir. 1955. "On Translating 'Eugene Onegin.'" *The New Yorker*, January 8, p. 34.

Nabokov, Vladimir. 1990. *Strong Opinions*. New York: Vintage Books (a division of Random House, Inc.). First Vintage International Edition, January, 1990. Originally published, in hardcover, by McGraw-Hill Book Company, New York, in 1973.

Nabokov, Vladimir. 1999. *Speak, Memory*. New York: Alfred A. Knopf (a division of Random House). First Everyman's Library Edition.

Peris, Álvaro, Miguel Domingo, and Francisco Casacuberta. 2016. "Interactive neural machine translation." *Computer Speech and Language* 45, 201–220.

Pushkin, Alexander. 1995. *Eugene Onegin: A Novel in Verse*. James E. Falen, translator. New York: Oxford University Press.

Pushkin, Alexander. 2002. *Eugene Onegin: A Novel in Verse*. Walter Arndt, translator. Woodstock: Ardis Publishers.

Pushkin, Alexander. 1979. *Eugene Onegin: A Novel in Verse*. Charles H. Johnston, translator. New York: Penguin Putnam.

Pushkin, Alexander. 2016. *Yevgeny Onegin*. Anthony Briggs, translator. London: Pushkin Press.

Pushkin, Alexandr. 1964. *Eugene Onegin: A Novel in Verse*. Vladimir Nabokov, translator. Princeton, NJ: Princeton University Press, Bollingen Series LXXII.

Seligman, Mark. 1991. *Generating Discourses from Networks Using an Inheritance-Based Grammar*. Dissertation. Berkeley: Department of Linguistics, University of California.

Seligman, Mark. 2000. "Nine Issues in Speech Translation." *Machine Translation*, 15(1/2), 149–186. Special Issue on Spoken Language Translation.

Seligman, Mark. 2019. "The evolving treatment of semantics in machine translation." In *Advances in Empirical Translation Studies*, Christine Ji, ed. Cambridge: Cambridge University Press.

Seligman, Mark and Alexander Waibel. 2019. "Advances in speech-to-speech translation technologies." In *Advances in Empirical Translation Studies*, Christine Ji, ed. Cambridge: Cambridge University Press.

Turney, Peter D. and Patrick Pantel. 2010. "From frequency to meaning: Vector space models of semantics." *Journal of Artificial Intelligence Research* 37 (2010), pp. 141–188.

Uchida, Hiroshi. 1986. "Fujitsu machine translation system: ATLAS." In *Future Generation Computer Systems*, 2 (2), pp. 95–100.

Wilson, Edmund. 1965. "The strange case of Pushkin and Nabokov." *The New York Review*, July 15.

Woszczyna, Monika, Matthew Broadhead, Donna Gates, Marsal Gavaldà, Alon Lavie, Lori Levin, and Alex Waibel. 1998. "A modular approach to spoken language translation for large domains." In *Proceedings of the Third Conference of the Association for Machine Translation in the Americas (AMTA) 98*. Langhorne, PA, October 28–31.

Appendix I

Pushkin variants

Chapter II: Aging

Oliver Elton/A.D.P. Briggs

When to the standard we are flying
Of tranquil reason, and her rule,
And when our passions' flame is dying
And we begin to ridicule
Their wilfulness and all their sallies
And their belated after-rallies,
Then, with a struggle, we are tame;
But sometimes like to hear the same
Wild speech of passion, in a stranger;
It stirs our heartstrings. So, while penned
In his forgotten hut, may lend
An eager ear to tales of danger
Some crippled veteran, when they're told
By young, mustachioed heroes bold.

James Falen

When we at last turn into sages
And flock to tranquil wisdom's crest;
When passion's flame no longer rages,
And all the yearnings in our breast,
The wayward fits, the final surges,
Have all become mere comic urges,
And pain has made us humble men –
We sometimes like to listen then
As others tell of passions swelling;
They stir our hearts and fan the flame.
Just so a soldier, old and lame,
Forgotten in his wretched dwelling,
Will strain to hear with bated breath
The youngbloods' yarns of courting death.

Vladimir Nabokov

When we have flocked under the banner
of sage tranquility,
when the flame of the passions has
    gone out
and laughable become to us
their willfulness, [their] surgings
and tardy repercussions,
not without difficulty tamed,
sometimes we like to listen
to the tumultuous language of another's
    passions,
and it excites our heart;
exactly thus an old disabled soldier
does willingly bend an assiduous ear

Charles Johnston

When we've retreated to the banner
of calm and reason, when the flame
of passion's out, and its whole manner
become a joke to us, its game,
its wayward tricks, its violent surging,
its echoes, its belated urging,
reduced to sense, not without pain –
we sometimes like to hear again
passion's rough language talked by others,
and feel once more emotion's ban.
So a disabled soldier-man,
retired, forgotten by his brothers,
in his small shack, will listen well
to tales that young mustachios tell.

Walter Arndt

When we have rallied to the standard
Of a well-tempered quietude,
And blazing passions have been rendered
Absurd, their afterglow subdued,
Their lawless gusts and their belated
Last echoes finally abated –
Not without cost at peace again,
We like to listen now and then
To alien passion's rage and seething,
And feel its clamor at our heart;
We play the battered veteran's part
Who strains to listen, barely breathing,
To exploits of heroic youth,
Forgotten in his humble booth.

Douglas Hofstadter

When once we've hoist the flag of aging
Rational men of mind serene,
And once the flame's been snuffed
    of raging
Passion (amen!), our old routine
Seems quaint and droll: those
    stubborn yearnings,
Those outbursts, and those mid-life
    churnings,
Though it took time, at last
    we're tame;
We savor now a gentler game:
Vicarious pangs of youthful tension,
For oft they'll touch our very core.

to the yarns of young mustached braves,
forgotten in his shack.

Just so, a grizzled man of war
Will crane his neck with rapt attention
To hear tales told, in his small shack,
By front-line johnnies just marched back.

Chapter III: Letter

Oliver Elton/A.D.P. Briggs

Tatyana's letter never tires me
To read; and when I read it now,
I hold it sacred; it inspires me
With a sad, private pang, I vow.
Who taught her in soft words to render
Her love, so heedless and so tender?
Such touching nonsense – to impart
All the wild language of her heart,
So baneful in its fascination.
I know not – a pale copy give,
No more – the picture does not live –
A feeble, incomplete translation;
Just so a schoolgirl's finger may,
All timidly, *Der Freischütz* play …

Charles Johnston

Tatyana s letter, treasured ever
as sacred, lies before me still.
I read with secret pain, and never
can read enough to get my fill.
Who taught her an address so tender,
such careless language of surrender?
Who taught her all this mad, slapdash,
heartfelt, imploring, touching trash
fraught with enticement and disaster?
It baffles me. But I'll repeat
here a weak version, incomplete,
pale transcript of a vivid master,
or *Freischütz* as it might be played
by nervous hands of a schoolmaid.

James Falen

Tatyana's letter lies beside me,
And reverently I guard it still;
I read it with an ache inside me
And cannot ever read my fill.
Who taught her then this soft surrender,
This careless gift for waxing tender,
This touching whimsy free of art,
This raving discourse of the heart –
Enchanting, yet so fraught with trouble?
I'll never know. But none the less,
I'll give it here in feeble dress:
A living picture's pallid double,
Or *Freischütz* played with timid skill
By fingers that are learning still.

Walter Arndt

What Tanya wrote is in my keeping,
I treasure it like Holy Writ;
I cannot read it without weeping
Nor ever read my fill of it.
Who, what, unsealed that fount of feeling,
With such unguarded grace revealing
(Naïve appeal of artless art)
Her unpremeditating heart,
Alike disarming and imprudent?
I cannot answer – anyhow,
Here is my weak translation now,
Life's pallid copy by a student,
Or *Freischütz* waverlingly played
By pupils awkward and afraid.

Vladimir Nabokov

Tatiana's letter is before me;
religiously I keep it;
I read it with a secret heartache
and cannot get my fill of reading it.
Who taught her both this tenderness
and amiable carelessness of words?
Who taught her all that touching [tosh],
mad conversation of the heart
both fascinating and injurious?
I cannot understand. But here's
an incomplete, feeble translation,
the pallid copy of a vivid picture,
or *Freischütz* executed
by timid female learners' fingers.

Douglas Hofstadter

Tatyana's missive lies before me;
To it religiously I cling.
Each time I read it, secret stormy
Sensations storm me, stir me, sting.
Who instilled in her this graciousness,
Tender, careless, strange loquaciousness?
Who taught her tongue to make no sense,
Her heart to rave with no pretense?
Such candid bubbling's sweet but risky.
Her source I cannot guess; but read
This version, pale and flat – indeed,
To what she penned as ale's to whiskey,
Or, one might say, *Freischütz* performed
By timid fingers still unwarmed:

Appendix II

Marot variants

Clément Marot	*Hofstadter 2b (literal)*	*Hofstadter 6b*
A une Damoyselle malade	**To a Sick Damsel**	**My Sweet Dear**

Ma mignonne,	*My sweet,*	*My sweet dear,*
Je vous donne	*I bid you*	*I send cheer –*
Le bon jour;	*A good day;*	*All the best!*
Le séjour	*The stay*	*Your forced rest*
C'est prison.	*Is prison.*	*Is like jail.*
Guérison	*Health*	*So don't ail*
Recouvrez,	*Recover,*	*Very long.*
Puis ouvrez	*Then open*	*Just get strong –*
Votre porte	*Your door,*	*Go outside,*
Et qu'on sorte	*And go out*	*Take a ride!*
Vitement,	*Quickly,*	*Do it quick,*
Car Clément	*For Clément*	*Stay not sick –*
Le vous mande.	*Tells you to.*	*Ban your ache,*
Va, friande	*Go, indulger*	*For my sake!*
De ta bouche,	*Of thy mouth,*	*Buttered bread*
Qui se couche	*Lying abed*	*While in bed*
En danger,	*In danger,*	*Makes a mess,*
Pour manger	*Off to eat*	*So unless*
Confitures;	*Fruit preserves;*	*You would choose*
Si tu dures	*If thou stay'st*	*That bad news,*
Trop malade,	*Too sick,*	*I suggest*
Couleur fade	*Pale shade*	*That you'd best*
Tu prendras,	*Thou wilt acquire,*	*Soon arise,*
Et perdras	*And wilt lose*	*So your eyes*
L'embonpoint.	*Thy plump form.*	*Will not glaze.*
Dieu te doint	*God grant thee*	*Douglas prays*
Santé bonne,	*Good health,*	*Health be near,*
Ma mignonne.	*My sweet.*	*My sweet dear.*

Hofstadter 43b	*Robert French II 9*
Goldilocks	**Fairest Friend**

Goldilocks,	*Fairest friend,*
Feisty fox,	*Let me send*
You're a pip,	*My embrace.*
Whom the grippe,	*Quit this place,*
Sad to say,	*Its dark halls*
Has in sway.	*And dank walls.*
Gotta fight!	*In soft stealth,*
With a right	*Regain health:*
To the chin,	*Dress and flee*

Hofstadter 43b

Babe, you'll win!
No kid gloves!
Clement loves
You, ya vamp –
You're his champ!
Champs must eat;
Wimpy wheat
Bread's a sham,
Without jam!
To gain brawn,
Champs chomp on
Jelly dough-
Nuts; they go
Nuts for pies
(Your top prize-
fighters do).
As for you,
Box that pox,
Goldilocks!

Robert French II 9

off with me,
Clement, who
Calls for you.
Fine gourmet,
Hid from day,
Danger's past,
So at last
Let's be gone,
To dine on
Honeyed ham
And sweet jam.
If you're still
Wan and ill,
You will cede
Pounds you need.
May God's wealth
Bless your health
Till the end,
Fairest friend.

6 A prototype system for multilingual data discovery of International Long-Term Ecological Research (ILTER) Network data

Kristin Vanderbilt, John H. Porter,
Sheng-Shan Lu, Nic Bertrand, David Blankman,
Xuebing Guo, Honglin He, Don Henshaw,
Karpjoo Jeong, Eun-Shik Kim, Chau-Chin Lin,
Margaret O'Brien, Takeshi Osawa, Éamonn Ó
Tuama, Wen Su, and Haibo Yang

6.1 Introduction

The International Long-Term Ecological Research (ILTER) Network, consisting of site-based research networks in 40 countries, collects long-term research and monitoring data from many ecosystems around the globe. Since its inception in 1993, this "network of networks" has collected a wide variety of data at its 633 sites (Figure 6.1). The aim of the ILTER is to contribute to the understanding of international ecological and socio-economic issues through the synthesis of data at broad temporal and spatial scales that may span multiple countries (Vihervaara et al., 2013; Haase et al., 2016). One barrier to compiling datasets to explore data from more than one country is the multilingual nature of the ILTER's data archives (Vanderbilt et al., 2010, 2015). Each national network manages its data using its own local language. This poses a difficulty for scientists seeking data outside of their own national network.

Successful sharing of data and information in the ILTER requires a common language that imparts understanding of what the data mean, as well as tools to do cross-language information retrieval. One tool that can be used to help facilitate data discovery is a thesaurus. A thesaurus is a structured and organised set of terms, usually about a specific domain, that can be used to index datasets or documents so that end-users can retrieve relevant information when searching using those terms (Broughton, 2006). Thesaurus terms are cross-referenced to other terms in the thesaurus that may be equivalent (synonyms), narrower than, broader than, or related to the term (Figure 6.2) (Clarke, 2001). This

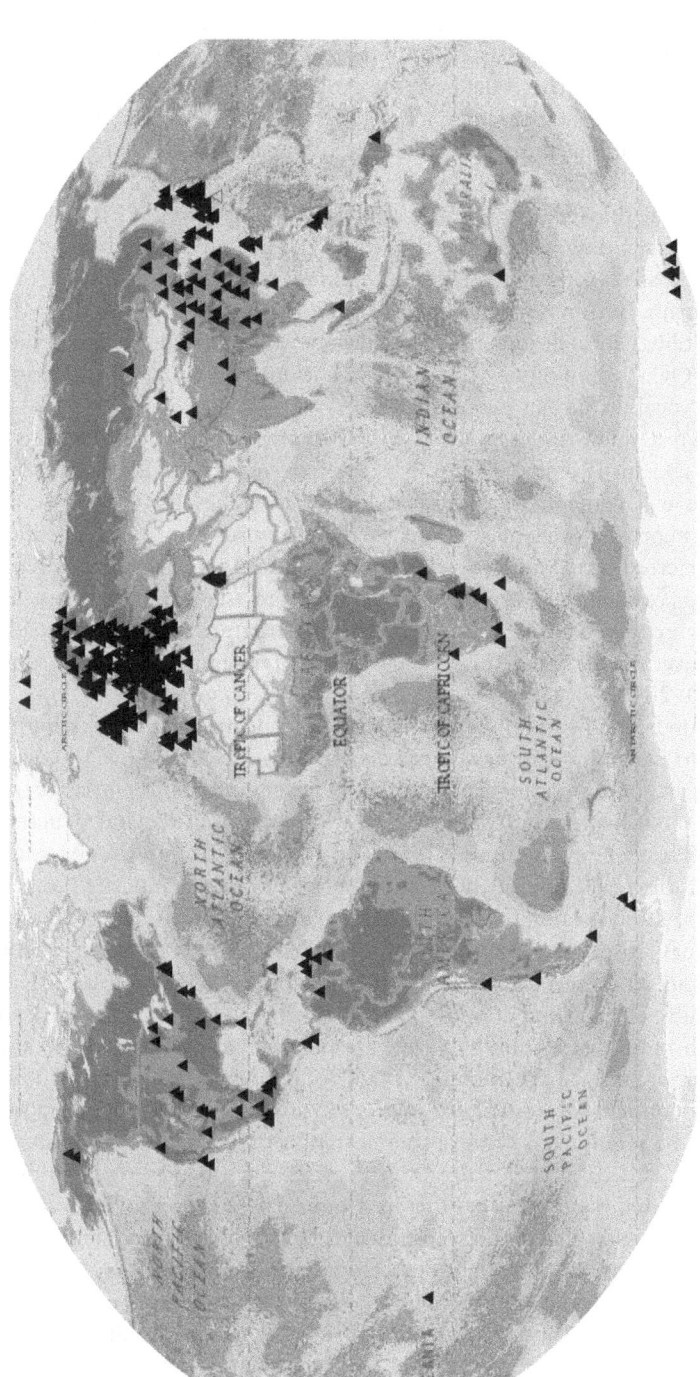

▲ ILTER Research Locations

Content may not reflect National Geographics current map policy. Sources: National Geographic, Esri, DeLorme, HERE, UNEP-WCMC, USGS, NASA, ESA, METI, NRCAN, GEBCO, NOAA, increment P Corp.

Figure 6.1 International Long-Term Ecological Research (ILTER) Network research site locations.

structure serves as a navigational aid to an end-user, placing terms in a hierarchical context and alerting the user to related terms to search with. A thesaurus also constrains the terms that a data creator can choose from as they select suitable terms to describe their documents or datasets. Both the data creator and end-user benefit from having a controlled list of vocabulary terms from which to select.

A monolingual thesaurus is useful within a single national LTER network, but to facilitate data discovery across the whole ILTER network, adoption of a multilingual thesaurus is needed. Several multilingual thesauri exist for the environmental domain, but they are too broad for use by the ILTER (e.g., GEMET [General Multilingual Environmental Thesaurus; www.eionet.europa.eu/gemet] and AGROVOC [Multilingual Agricultural Thesaurus; http://aims.fao.org/vest-registry/vocabularies/agrovoc]).

Even within a single monolingual LTER Network, creating a thesaurus is a challenge. Thesaurus creators must first select terms to include in the thesaurus. These will come from published lists, dictionaries, databases, or the collection of items that will be indexed by the thesaurus (Broughton, 2006). Then, the preferred term must be selected from synonyms or spelling variants (e.g., color vs. colour), and the terms organised into a hierarchical structure. Related terms are then organised into a hierarchical structure specifying "broader than," "narrower than," "related to," and "use for" relationships between terms (ANSI/NISO, 2010).

Methods for creating a multilingual thesaurus include merging existing monolingual thesauri, starting with a new thesaurus and considering multiple languages from the outset, or translating an existing thesaurus into multiple languages (IFLA, 2009). No matter the approach taken, term equivalence and structural challenges will likely be encountered (Jorna and Davies, 2001). In the context of a multilingual thesaurus, equivalent terms should be both semantically (i.e., the terms have the same meaning) and culturally equivalent (IFLA, 2009). Partial equivalence may arise when a term in one language has a somewhat broader or narrower meaning than a term in another language, or the translated term may have a different cultural connotation. The terms "loud" and "noisy," for instance, both mean "easily audible," but are only partially equivalent because "noisy" has a more negative connotation than "loud." An equivalent term in one language may not exist for a particular concept in another, and two terms in one language may be required to capture the meaning of the preferred term in the other. Semantic and cultural differences in the use of terms may result in non-symmetrical hierarchies of terms in different languages. However, one advantage to using a multilingual thesaurus, rather than a simple list of translated words, is that concepts that may be ambiguous or difficult at one level may be direct translations at another level in the hierarchy. For example, Vanderbilt et al. (2010) showed how the Japanese and English concepts for "wetlands" are different, but many lower-level units (different types of wetlands such as salt marsh or mangrove) are similar. So searches on the high-level term will still find data tagged with the more specific terms.

Net Primary Productivity
    BT: Primary Productivity
    RT: Carbon Dioxide
        UF. CO2
    RT: Trophic Levels
        UF: Energy Levels
        BT: Food Chains
        RT: Feeding Habits
        RT: Saprophytism
        RT: Primary Productivity
        RT: Biological Production
        RT: Net Primary Activity

Figure 6.2 An excerpt from AGROVOC illustrating the hierarchical nature of a thesaurus. Descriptors mean: BT: broader than; NT: narrower term; RT: related term; UF: used for. For a data creator designating keywords for a dataset, the thesaurus would tell them to use the term "carbon dioxide" instead of CO_2. An end-user searching for data indexed with the term "Primary Productivity" would retrieve records tagged with "Trophic Levels" as well, if the query engine is set to return "related terms."

Ideally, an ILTER end-user would query the ILTER data archive using a term in their own local language and be able to retrieve resources tagged with that term in other languages in the database. To accomplish this query, software is needed that can use the multilingual thesaurus to find translations and then query stores of ILTER data using the translated terms. The adoption of a common software stack and metadata standard for managing data in many ILTER national networks makes this task tractable. In 2010, ILTER members agreed to use Ecological Metadata Language (EML) (Fegraus et al., 2005) as the metadata standard for the network (Vanderbilt et al., 2010). EML is implemented as a set of XML schemas that can be used in an extensible manner to document ecological data. Metacat (Berkley et al., 2001), a database for storing data packages (i.e., data + metadata) is an open-source solution for managing data and metadata and many ILTER network members use it (e.g., Lin et al., 2008; Ohte et al., 2012). Metacat stores XML documents in a relational database, from which they can be queried using a path-oriented query language. Metacat will store metadata documents in different languages. ILTER data are not stored in one centralised location, but in a distributed system of Metacats.

In 2012, the "Semantic Approaches to Discovery of Multilingual ILTER Data" workshop was held at the East China Normal University in Shanghai, China to explore how to improve data discoverability in the multilingual ILTER data archives. The results of that workshop were wide-ranging, varying from evaluation of search resources, enhancement of existing thesauri and development of a prototype distributed ecological data system. Here we describe the steps required to build a prototype web-service-based multilingual data search system, including development of base thesauri, both monolingual and multilingual, development of interfaces and search tools and subjective assessments of the prototype multilingual data search system. We also discuss lessons learned from the prototype that can be used to guide creation of a production system, and the degree to which web-based automated translations might be used to make full metadata translations available.

6.2 Building a monolingual thesaurus

An example of the process of building a thesaurus comes from the US LTER Network. Historically, most keywords used to characterise datasets at US LTER sites were uncontrolled. They were selected entirely by the data creator without reference to words used in other datasets. One of the challenges facing researchers in discovering data from LTER sites was inconsistent application of keywords. A researcher interested in carbon dioxide measurements would need to search on both "Carbon Dioxide" and "CO2." Moreover, the existing set of keywords was highly diverse. For example, in a 2006 survey of Ecological Metadata Language (EML) documents in the LTER Data Catalog, over half the keywords (1616 of 3206) were used in only a single dataset, and only 104 (3%) of the keywords were used at five or more different LTER sites (Porter, 2006; Porter and Costa, 2006).

To address this problem, in 2005 the US LTER Information Management Committee (comprised of one information manager from each of the 26 US LTER sites) established an ad hoc "Controlled Vocabulary Working Group" and charged it with studying the problem and proposing solutions. The group compiled and analysed keywords found in LTER datasets and documents, and identified external lexicographical resources, such as controlled vocabularies, the-sauri and ontologies, that might be applied to the problem (Porter, 2010). Initially the working group attempted to identify existing resources, such as the GEMET Thesaurus, the Global Change Master Directory keyword list, and the National Biological Information Infrastructure (NBII) Thesaurus (now the US Geological Survey Biocomplexity Thesaurus), that LTER might be able to adopt wholesale.

Unfortunately, using matches with widely-used LTER keywords as a metric, none of the external resources proved to be suitable. Too many keywords commonly used in LTER datasets were absent from the existing lexicographical resources. So, starting in 2008 the working group focused on developing an LTER specific controlled vocabulary, ultimately identifying a list of approximately 600 keywords that were either used by two or more LTER sites, or were found in one of the external resources (NBII Thesaurus and Global Change Master Directory Keyword List). Excluded from the list were taxonomic names for species and names of geographic locations, as these were considered to be better addressed using existing taxonomic resources and gazetteers. The form of keywords were adjusted to conform to the recommendations of the international standard for controlled vocabularies (ANSI/NISO, 2010), but the original forms were preserved as synonyms or "use for" terms to facilitate searching. This draft list was then circulated to members of the US LTER Information Management Committee for suggested additions and deletions, which were then voted upon (Porter, 2010).

Organisation of the keywords into a polytaxonomy (i.e., multiple taxonomies) and thesaurus followed the recommendations of the "Guidelines for the Construction, Format, and Management of Monolingual Controlled Vocabularies" (ANSI/NISO, 2010). Members of the Controlled Vocabulary Working Group classified each keyword into one of six different types (things, properties, processes, materials, disciplines, and events). This greatly simplified the organisational process by allowing them to focus on a smaller subset of terms when organising them into a hierarchical structure, or taxonomy. Using the Tematres online thesaurus software (www.vocabularyserver.com/index. html), ten taxonomies were created. Four taxonomies were of type "things" (Organisms, Ecosystems, Organizational units, and Substrates), two taxonomies were of type "processes" (Processes, Methods) and the other four taxonomies (Substances, Measurements, Events, and Disciplines) were each of one of the four remaining types. Some additional terms were added to facilitate grouping (e.g., "hydrologic properties") and some terms that were found to be too ambiguous when used alone (e.g., "aboveground") were deleted. Synonyms, abbreviations, variant spellings were added as "use for" terms to facilitate searching of existing metadata documents that had not yet been revised to incorporate preferred terms. Version 1.0 of the US LTER Controlled Vocabulary Working Group Thesaurus contained 627 preferred terms and an additional 150 "use for" terms (http://vocab.lternet.edu, Porter, 2010).

6.3 Creating a multilingual thesaurus

EnvThes is a thesaurus developed by European projects EnvEurope (www.enveurope.eu/) and ExpeER (http://expeeronline.eu/). It provides a common set of defined concepts that can be used to annotate the heterogeneous data collected and managed in different ways at research sites throughout Europe. It was selected for use by the ILTER because it is the most comprehensive list of terms available to describe LTER activities of ecological monitoring, research, and experiments.

EnvThes is constructed from several existing vocabularies and thesauri, including the US LTER Thesaurus. To create a comprehensive list of concepts covering the wide range of disciplines studied by the European ecological community, terms from the INSPIRE spatial data themes (http://inspire.ec.europa.eu/index.cfm/pageid/2/list/7), EUNIS habitat types (http://eunis.eea.europa.eu/habitats.jsp), and NASA units controlled vocabularies were included (Schentz et al., 2013). English was established as the main language of EnvThes, and translations of concept definitions were made to provide multilingualism. English terms may or may not have translations, depending on the resources available for translation and the degree of equivalence of the translated term. EnvThes terms are linked to definitions in existing vocabularies such as AGROVOC, Wikipedia (www.wikipedia.org), EUROVOC (https://eur-lex.europa.eu/browse/eurovoc.html), GEMET (www.eionet.europa.eu/gemet/), and EARTh (https://vocabularyserver.com/cnr/ml/earth/en/index.php), so that EnvThes is part of a web of interlinked thesauri. Domain scientists do the translation of the concept definitions. As was the US LTER Thesaurus, EnvThes is implemented using the Simple Knowledge Organization System (SKOS; Miles and Bechhofer, 2009) which allows these links to indicate "exact match," "close match," "narrow match," "broader match" and "related match." This system of linked thesauri creates a stable semantic reference for terms in EnvThes.

EnvThes, at the time the 2012 workshop, had terms translated into fourteen European languages. Additional languages needed to be added, because thirty-three languages are used in the ILTER. During the workshop, Korean, Japanese, and Chinese participants contributed to the work by translating the 627 terms from the US LTER Thesaurus that are in EnvThes. This was readily achieved during the workshop, although some of the English terms did not have exact equivalents in the target language. Regional biases for terms were noted, e.g., "typhoon" is the English word used for "hurricane" in Asia, but in EnvThes the preferred term is "hurricane" which reflects the preferred term in the US LTER Thesaurus.

6.4 Multilingual search prototype

6.4.1 Implementation

A web-services framework was adopted for development of the prototype multi-lingual ecological data search engine. Use of web services facilitated the use of existing systems and eliminated the need to duplicate existing functionality. EnvThes is currently implemented using the Resource Description Framework (RDF)-based TopBraid software (www.topquadrant.com/) which incorporates a SPARQL endpoint for remote access. Similarly, the Metacat data catalogs incorporate a web-services application programming interface (API) that allow queries to be submitted as XML documents, with the results returned as XML as well. For ease in integration with the web, web services to perform the needed functions were written in the PHP language. The ARC2 PHP module was used to facilitate communication with an EnvThes instance established to support the prototype using SPARQL.

Implementation of the prototype thesaurus-enhanced multilingual search interface required several components (Figure 6.3):

- Preferred-term autocomplete – Helps guide users to existing terms
- Thesaurus – Needed to make connections between terms
- Search interface – Web page from which users select terms to search in their local language
- Enhancer – Uses the underlying Thesaurus to generate an expanded list of search terms by including synonyms and narrower or related terms to the user-provided term
- Translator – Translates the original search term and expanded terms into the language needed for the search engine

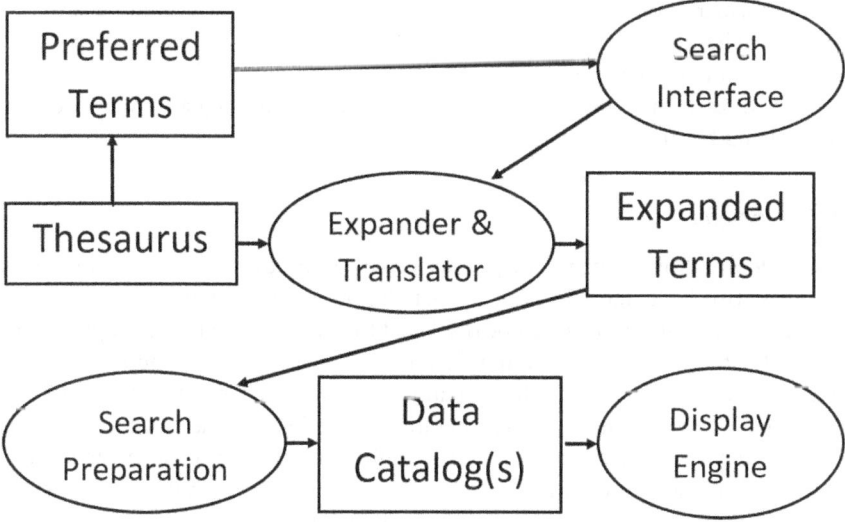

Figure 6.3 Diagram of multilingual search system elements.

- Search preparation – Formats the translated and enhanced terms into a query suitable for use with a specific data catalog or search engine
- Data catalog(s) – Stores the metadata, searches based on the prepared query, and returns a list of matching datasets
- Display engine – Translates the list of matching datasets into an attractive, user-readable form.

In the prototype, the process of identifying the initial preferred term was facilitated in the online search form by using an "autocomplete" list that suggested keywords from the thesaurus when a user started to type in a search string. This helped reduce "misses" caused by misspellings or selection of words not in the thesaurus. However, we did not implement this function for languages that use ideograms, such as Chinese, Japanese and Korean. For those languages a browse-based system may be preferable over an autocomplete system.

In our case, we were able to combine the Enhancer and the Translator functions by using the EnvThes Multilingual Thesaurus. In that thesaurus each preferred term is linked to a concept. Each concept has multiple alternative labels in different languages. Thus, a list of keywords for an enhanced multilingual search can be created by identifying a preferred term to serve as a starting point (e.g., "aquatic ecosystems"). Synonyms and narrower and/or related terms can then be identified and alternate labels in a particular language extracted (e.g., in French: "Lac," "Rivière," and "Ruisseau"). Once a list of terms has been extracted in the desired language, they need to be formatted into a form suitable for use in a search engine. In our prototype application, our targets were all Metacat data catalogs (Jones et al., 2001). These supported a web service interface that used a format known as "path Query" which specified what should be searched and how elements of a search should be combined. In our case, we simply used a "union" operator to return datasets that had any one of the keywords in the enhanced search in it. The resulting list was reformatted using an XML stylesheet for viewing by the user.

6.4.2 Interface

A sample workflow shows the web pages seen by users during a search (Figure 6.4). It starts with language selection, so that the proper autocomplete list of words in the user's local language can be displayed for the user and the appropriate language for the final search selected. It then proceeds to obtain the search term. The prototype takes the search term and translates and expands that term using multiple calls to the multilingual thesaurus. Multiple calls are needed because the thesaurus software typically returns only a single level of a hierarchy at a time, so burrowing down multiple levels requires multiple calls. In the prototype, resultant sets of search terms, ordered by complexity (e.g., synonyms only, synonyms and narrower terms, synonyms, narrower terms and related terms) are provided for the user to view.

However, this step could be omitted for a production system and the search at a pre-determined level of complexity run automatically. The user is then queried for the data catalog to search and the search results are displayed. Each data catalog typically contains metadata only in the local language, although some may support more than one language. Translations are made into only a single language, so in the case where the Metacat contains more than one language, the user chooses which language to query.

A potential enhancement to the system would be to automate searches of multiple data catalogs, each in their preferred language. However, this would require that the enhancement/translation step be repeated multiple times and that the results returned from the different data catalogs be integrated, with any duplicates removed. The web services approach used in this prototype system would also allow non-Metacat data catalogs to be queried. For instance, the Chinese Ecosystem Research Network (CERN) does not use Metacat or EML but could still use web services to receive queries translated into Chinese from the prototype system and then return target datasets.

6.4.3 Evaluation

The prototype multilingual search application was effective in improving most searches. By automatically expanding the number of search terms to include synonyms and narrower terms, most test searches were able to return relevant datasets. However, there were some limitations apparent in the prototype. First the number of translated terms varied widely across languages within the thesaurus. English had the largest number of terms in the EnvThes thesaurus used

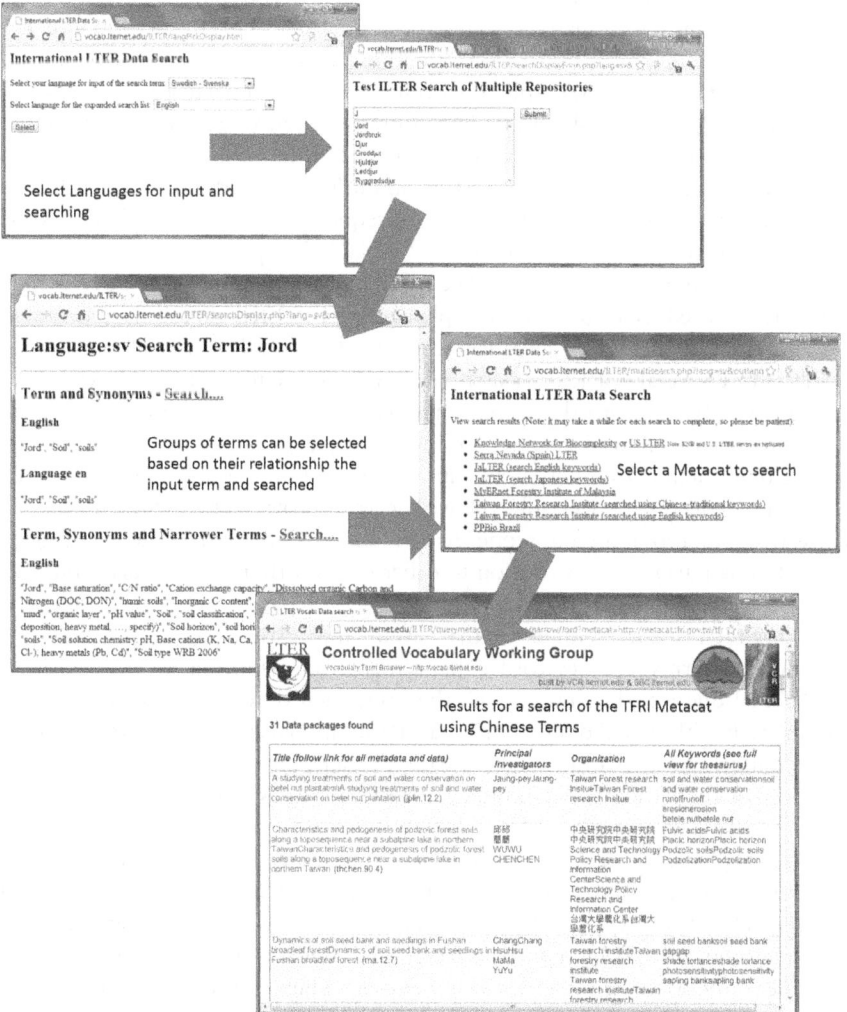

Figure 6.4 A prototype system for multilingual searching of ecological data. The user selects the language they wish to use for search input (e.g., Swedish) and the Metacat to be searched (e.g., Taiwan Forestry Research Institute (TFRI) Metacat). A term autocomplete list in Swedish guides the searcher towards terms in the thesaurus. The thesaurus is used to select additional terms such as synonyms or narrower terms prior to preparing a search. The enhanced search "path Query" is then sent to the TFRI Metacat after being translated into the language appropriate for that particular Metacat (Chinese in this case).

at the workshop, with over 1000 terms. In contrast many other languages were represented by 100 terms or fewer and a few, such as traditional Chinese, were represented by fewer than 10. This meant that the effectiveness of the keyword enhancement for translation into those languages was relatively limited. However, effectiveness was still enhanced if the user-supplied search term could be translated into English and the resulting set of terms extracted from the thesaurus could be used. The translation of over 600 terms into Japanese, Chinese (simple), Chinese (traditional) and Korean during the workshop should dramatically help to remedy this limitation in subsequent versions of the EnvThes thesaurus, at least for Asian languages.

A second limitation was speed. The prototype multilingual data search was too slow to be used as a production implementation. Often several minutes might be required at the search enhancement and data search stages of the process. There were several bottlenecks associated with the system. First, the system was highly distributed, with the search interface in the United States, the thesaurus in Europe and the data catalogs in Taiwan, Japan, Spain, Brazil, Malaysia and the United States. Internet latency in any of the long-distance links caused the system to slow down. Secondly, the web services provided by the thesaurus were effective, but also limited in scope. As noted above, any term could be extracted based on language and its relation to other terms within the thesaurus, but each level traversed in the thesaurus required that a separate web service query be sent for each of the members at that level. Thus, if the search term "soil" had 10 "child" terms, the keyword enhancement tool would need to perform 11 individual queries (the first on soil, then one each on each of the child terms). But if each of those child terms each had 10 additional children, the total number of required queries would be 111, each of them involving an intercontinental trip across the network. Additionally, for the purposes of testing, the prototype tool created alternative sets of search terms, using different rules regarding what should be included (e.g., synonyms only, synonyms and narrower, synonyms, narrower and related, and synonyms, narrower, related and the narrower terms of the related terms) that the user could select. This required that the entire enhancement process be repeated several times. Finally, the "path Query" search engine in the Metacat server is slow, especially when confronted with a large or complex query, such as those generated by the search enhancement web service. Complex searches could take up to several minutes to return a result.

Most of these performance difficulties can be relatively easily addressed in a production system. Eliminating the multiple search options (i.e., only searching for synonyms and narrower terms) is one obvious way to speed up the system. Similarly, moving the query-intensive functions closer to the server being queried would be a substantial help in reducing network latency. If necessary, applications that operate directly on the thesaurus database, rather than indirectly through a web service, could be used to reduce the time spent processing queries. If a single query sent to the thesaurus server could return the full hierarchy of needed terms instead of only a single level, speed would be substantially enhanced. Additionally, upgrades in the web service support in the component software could speed up searches. For example, recent versions of the Metacat software have added the capability to use the extremely fast Apache Solr search engine (Shahi, 2015), in place of the very general, but also very slow, "path Query" search. Finally, for a production system, periodic caching of either enhanced search term lists or even caching of data search results could drastically improve performance.

6.5 Automated translation of metadata

The multilingual search tool can help users locate suitable data, but that data will still not be useful if researchers cannot correctly interpret the metadata itself. It is beyond the scope of the prototype multilingual search system to provide full translations of the metadata. However, there are a variety of web-accessible translation engines that might be applied to this task. To evaluate the adequacy of those engines, we extracted the abstract from an EML metadata document and used several different web translation tools to transform it into English and compared that to the direct translation provided by the metadata author.

The original abstract was in Chinese (traditional):

本研究主要在建立觀霧地區大鹿林道鍬形蟲之物種名錄及族群數量等資料,自2002年4月至2003年3月間,於觀霧地區之大鹿林道設置樣站,以夜間燈光採集法、沿路調查法、誘餌採集法和朽木採集法等四種方法進行調查。本區全年共記錄14屬28種440隻鍬形蟲,其中以高砂深山鍬形蟲　(Lucanus taiwanus) 為最優勢的種類,佔全部記錄數量的13.6%。

The translations of this paragraph into English using three online translation tools were:

6.5.1 Google Translate Version http://translate.google.com/

This study aimed to establish Guanwu area Dalu Forest Road stag insects of the species directory and ethnic groups number and other information from April 2002 to March 2003, in view of the fog area of the large deer forest roads set the kind of station, the night light collection method along survey, bait collection and deadwood collection of four methods of investigation. This year, recorded a total of 14 genera of 28 kinds, 440, stag beetles, including high sand mountains stag beetles (Lucanus taiwanus) was the most dominant species, accounting for 13.6% of all records.

6.5.2 Bing Translate Version www.microsofttranslator.com/

This study mainly in view of establishing fog area lulindao rohac of list and data such as numbers of species, from April 2002 to March 2003, in view of fog district of lulindao station, along to night-time lighting collection law, investigation, decoy collection and Deadwood collection law, four kinds of methods to investigate. This area a total of 28 species of the record 14 440 rohac, mountains with high sand shovel shaped beetle (Lucanus taiwanus) for most types of advantages, 13.6% percent of total recorded amount.

6.5.3 World Lingo Version www.worldlingo.com/en/ products_services/worldlingo_translator.html

This research mainly establishment view fog area moose wood-road spade shape insect in materials and so on species of name list and tribal grouping tribal group quantity, from April 2002 to March 2003 between, Yu Guanwu moose of wood-road establishment type station the area, by at night light gathering law, along the road the survey procedure, the bait gathering law and the rotten wood gathering law and so on four methods carried on the investigation. The home court whole year altogether records 14 to be 28 kind of 440 spade shape insect, in which take the high granulated substance remote mountain spade shape insect (Lucanus taiwanus) as the most superiority type, occupies records quantity completely 13.6%.

Finally, the author's original English Version was:

> The purpose of this study was to draw up a namelist of stag beetles with their abundances along the Da-lu Forest Road in the Kuanwu area of northwestern Taiwan. From April 2002 to March 2003, four different methods, including light traps, transect line sampling, bait sampling, and rotten wood chopping method, were used in the investigation. In total, 28 species belonging to 14 genera of Lucanidae with 440 of stag beetles were recorded during this investigation in the Kuanwu area. The most abundant species was Lucanus taiwanus which accounted for 13.6% of total individuals.

Although none of the translations successfully captured a fully intelligible version of all the methods used to census stag beetles, it is at least possible to understand enough of the abstract to determine whether the dataset might be useful. It is even likely that someone familiar with the different sampling approaches used to observe stag beetles would be able to correctly discern which methods were used, even if the terminology is not typical.

We also reversed the process. When the original English abstract was translated to Chinese, co-author Sheng-Shan Lu noted that the grammar was incorrect and some characters were not in the correct order. The collection methods were also not fully understandable, just as they had been in the Chinese to English translation. He estimates that he understood about 60% of the meaning of the translation and this was sufficient to allow a determination about its usefulness to him. A similar test was done for Japanese to English and vice versa, and the accuracy of translation was judged to be about 60% correct in both directions.

Translation success of English to Swedish and the reverse were also tested with Google Translate. As one would expect because Swedish and English are more closely related languages, automatic translations from Swedish to English or the reverse are quite good. The automatic translations were at least 90% semantically equivalent.

We did not attempt to replicate this translation experiment on all parts of the metadata. It is likely that the parts of the metadata relating to the actual structure of the underlying data tables should be even more intelligible. By combining translations of the metadata about the column headers and descriptions with the relevant units, a more nuanced understanding of the data in a column can be achieved. For example, if a measurement of "annual nitrogen deposition rate" is mistranslated as "nitrogen in year," the underlying unit of "grams per meter squared per year" or "g/m2/yr" should help to clarify any ambiguity.

6.6 Discussion

General ecological theories are best tested using data from widely disparate systems (Tonn et al., 1990; Brown, 1995). Ecological processes that are important at a local scale may be quite different from those at regional and global scales (Levin, 1992; Lawton, 1999; Gross et al., 2000). A common approach is to mine the ecological literature for data, but such data are often summarised or incomplete. Access to raw and voluminous ecological data via data repositories provides new opportunities (Michener and Jones, 2012), but also demand bridging language gaps in order to locate and access ecological data from different regions or continents.

Online searching for data even in a single language can be complicated. We found that for the US LTER sites, uncontrolled application of user-supplied keywords resulted in over 3000 keywords, the majority of which were used only once. The experience was much the same for the Taiwan data system. They found that almost 69% (990 of 1305) of the keywords were used only in a single dataset, and only 36 (2.8%) of the keywords were used at five or more times. Such a wealth of keyword diversity means that most searches based on a single keyword would result in only a single dataset, and that finding related datasets would require repeated searches of using different terms related to a subject. The situation is even more difficult in a multilingual context.

The Taiwan Forestry Research Institute found that the English and Chinese keywords from their Metacat were used often in uncontrolled and inconsistent ways, even by the same authors. The same concepts in English keywords differed in spelling or could be singular or plural. Inconsistencies were also found in the Chinese terms used that meant the same thing. As an example, the term "wireless sensor network (WSN)" was annotated inconsistently in Chinese. They found three different translations in Chinese for WSN, 無線感測網 (five Chinese characters), 無線感測網路 (six Chinese characters), 無線感應器網絡 (seven Chinese characters). The Chinese characters used vary from five to seven, but still had the same meaning.

There are two parallel approaches that can be used to help improve the reliability and efficiency of searches. One is to encourage, or require, the use of keywords drawn from a controlled vocabulary, thesaurus or ontology. Use of these preferred terms can help to remedy the "one keyword, one dataset" problem, because multiple datasets will share keywords. Moreover, it addresses the unnecessary and confusing variation caused by variant spellings or the use of plural vs singular. For new metadata, autocomplete forms or keyword browsers can help guide users to preferred terms when preparing metadata. However, for existing metadata this approach can be time-consuming and expensive, because it requires going back through existing metadata to standardise selection of keywords. However, typically keywords will still be in a single language, so translation problems remain.

The other approach, that we tested here, is to increase the intelligence of the search process through the use of a multilingual thesaurus. Inclusion of synonyms or "use for" terms in the thesaurus can address the issues associated with variant spellings or use of plural vs singular terms. Additionally, the structure available in a thesaurus, where more specific terms can be linked to broader "parent" terms, allows a much more complete and reliable search to be run. For example, a search on "forests" will also return datasets that include the narrower term "trees," even though the term "forest" is never mentioned in the metadata.

The tasks of translation can be moved from the metadata creation into the creation of the multilingual thesaurus, where each term need only be translated once. These two approaches are complementary. As the quality of metadata documents is improved by the inclusion of standardised, preferred terms, so is the quality of searches provided through use of the multilingual thesaurus. However, there are also other approaches, such as free text searches that do not depend on the identification of specific keywords.

The relative brevity of many metadata documents, relative to text documents written for more general purposes, may pose a challenge for free-text searches. Moreover, if the search does not utilise the context provided by the structural descriptors that are used to define different components of the metadata, it is likely to return many false results. For example, a search for the researcher with the surname "Young" may be confounded with all manner of youthful organisms, such as "young leaves" and "young of the year"; not to mention the popular wind sensor manufactured by the "R.M. Young Company." Searches that focus on particular elements of a metadata document, such as title and keywords, are less likely to make such mistakes.

Generally, studies have shown that use of a preferred term list can improve search precision (Mackenzie-Robb, 2010) over free-text searches. One of the primary advantages the multilingual thesaurus has over free-text search is in its accuracy of results. The use of preferred terms ensures that the meaning of terms is known and assures consistency in term use that can improve search performance. Use of a thesaurus can also reduce irrelevant returns from free-term searches that are often caused by the inherent ambiguity of natural language and incompatibilities in translation of these terms. Of course, to attain this improved accuracy will exact a price. Research projects will need to adopt the thesaurus as a means for assigning terms to data resources.

Limited term assignment will lead to unsatisfactory search results that miss relevant resources. Generation of multilingual thesauri is also challenging. In addition to the work required to generate a monolingual thesaurus, domain experts in each language will need to review and revise any automatic term and term definition translations to assure the term and its definition match the original concept. Additionally, terms for many languages will need to be manually translated. Beyond the work of creating a thesaurus, there are also management tasks, such as adding and defining new terms, deprecating outdated terms, and relating terms as needed.

Here we implemented enhanced searches using a single multilingual thesaurus, albeit one assembled from parts of existing thesauri. However, there is no reason that searches could not be enhanced using multiple thesauri, with synonyms and narrower terms drawn from several thesauri or ontologies. The main challenges would be developing the needed web services to harvest terms from each thesaurus and eliminating redundant terms. There would also be performance concerns, as the search could proceed only as fast as the slowest of the thesauri, and the larger number of search terms could slow down the search engines associated with data catalogs.

As discussed by Vanderbilt et al. (2010), multilingual ontologies, rather than thesauri, may offer the best long-term solution for facilitating sophisticated and accurate data discovery. Ontologies are models of concepts and their relationships within a scientific domain. Ontologies offer more relationship types (e.g., is-a, has-part, located-in) with which to capture semantic relationships between concepts (Madin et al., 2008). Work is ongoing to develop ontologies for the biological and ecological domains, in particular, the Biological Collections Ontology, the Environment Ontology, and the Population and Community Ontology (Walls et al., 2014). Vanderbilt et al. (2010) envisioned linking multiple monolingual ontologies with a core ontology (e.g., OBOE, Madin et al., 2007; SERONTO, van der Werf et al., 2008).

Such a structure would allow the wider array of relationship types available in ontologies to be exercised to further reduce the ambiguity inherent in multilingual searches. Development of thesauri pave the way for creation of ontologies (Almeida and Simoes, 2006; Rajbhandari and Keizer, 2012) but substantial additional effort to define both core ontologies and the monolingual ontologies is required. Additional effort is then required to link each of the ontologies into the core. For this reason Vanderbilt et al. (2010) recommended that thesaurus-based search be developed first, both to provide more immediate aid to searching and to help form the basis for the needed ontologies for a longer-term solution.

A more radical approach to providing access to multinational data is to retrieve actual data rather than metadata. This Linked Data approach combines computational ecology and eco-informatics (Gray, 2009) and refers to a set of best practices for publishing and interlinking structured or non-structured data on the web in a machine-readable way (Berners-Lee, 2006; Heath and Bizer, 2011; Wood et al., 2014). The Linked Data approach uses a general standard, the Resource Description Framework (RDF), to make data and metadata amenable to automated interpretation by computers. Although RDF provides a generic, graph-based data model encoding data in typed statements called triplets, it depends on the domain ontology or controlled vocabulary to specify the concepts and the relationships between concepts in the triplets (Bizer et al., 2009). Linked Open Data has been tested for ecological research in Taiwan using a monolingual data catalog (Mai et al., 2011) and it was found to be an effective way of sharing a wide variety of ecological data. The multilingual thesaurus developed by the ILTER is a first step towards building the sorts of ontologies needed to support linked data approaches in the future.

6.7 Conclusions

We found that a multilingual thesaurus-based data search capability could be developed using a web-service-based approach. There remain some issues regarding completeness of the thesaurus and performance, but those are issues that are readily addressed. We also found that processing metadata documents using existing online automated translation services, although far from completely accurate, provided enough information in most cases to support decisions regarding the suitability for use of specific datasets. There are additional and more sophisticated approaches that could also be used to support data discovery, but the thesaurus framework used here can serve as a useful stepping-stone, as well as a useful interim tool for researchers.

Acknowledgements

Funding: Travel support for US participants to the workshop in China was provided by a supplement grant to NSF DEB grant #021774. Local costs for all participants were generously covered by the Chinese Ecological Research Network (CERN) central office.

References

Almeida, J. J., Simoes, A., 2006. T2O – Recycling thesauri into a multilingual ontology. In: Calzolari, N., Choukri, K., Gangemi, A., Maegaard, B., Mariani, J., Odjik, J., Tapias, D. (Eds.), Proceedings of the Fifth International Conference on Language Resources and Evaluation (LREC), Genoa, Italy, May 22–28, pp. 1466–1471.
ANSI/NISO Z39.19-2005, 2010. Guidelines for the Construction, Format, and Management of Monolingual Controlled Vocabularies. NISO, Baltimore, Maryland.
Berkley, C., Jones, M., Bojilova, J., Higgins, D., 2001. Metacat: a schema-independent XML database system. In: Proceedings of the 13th International Conference on Scientific and Statistical Database Management, George Mason University, Virginia, July 18–20, IEEE Computer Society, Washington, DC, pp. 171–179.
Berners-Lee, T., 2006. Linked data – design issues. www.w3.org/DesignIssues/LinkedData.html (accessed March 17, 2015).
Bizer, C., Heath, T., Berners-Lee, T., 2009. Linked data – the story so far. Int. J. Semantic Web Inf. Syst. 5, 1–22.
Broughton, V., 2006. Essential Thesaurus Construction. Facet Publishing, London.

Brown, J. H., 1995. Macroecology. University of Chicago Press, Chicago, IL.

Clarke, S. G. D., 2001. Thesaural relationships. In: Bean, C. A., Green, R. (Eds.), Relationships in the Organization of Knowledge. Kluwer Academic Publishers, Boston, pp. 37–52.

Fegraus, E., Andelman, S., Jones, M. B., Schildhauer, M. P., 2005. Maximizing the value of ecological data with structured metadata: an introduction to the ecological metadata language (EML) and principles for metadata creation. Bull. Ecol. Soc. Am. 86, 158–168.

Gray, J., 2009. Jim Gray on eScience: a transformed scientific method. In: Hey, T., Tansley, S., Tolle, K. (Eds.), The Fourth Paradigm: Data-Intensive Scientific Discovery. Microsoft Research, Redmond, Washington, DC, pp. 1–16.

Gross, K. L., Willig, M. R., Gough, L., Inouye, R., Cox, S. B., 2000. Patterns of species density and productivity at different spatial scales in herbaceous plant communities. Oikos 89, 417–427.

Haase, P., Frenzel, M., Klotz, S., Musche, M., Stoll, S., 2016. The long-term ecological research (LTER) network: relevance, current status, future perspective and examples from marine, freshwater and terrestrial long-term observation. Ecol. Indic. 65, 1–3.

Heath, T., Bizer, C., 2011. Linked Data: Evolving the Web Into a Global Data Space. MC Publishers, San Rafael, CA.

IFLA Working Group on Guidelines for Multilingual Thesauri, 2009. Guidelines for Multilingual Thesauri, IFLA Professional Reports 115, International Federation of Library Association and Institutions (IFLA), The Hague, 30pp.

Jones, M. B., Berkley, C., Bojiova, J., Schildhauer, M., 2001. Managing scientific metadata. IEEE Internet Comput. 5, 59–68.

Jorna, K., Davies, S., 2001. Multilingual thesauri for the modern world – no ideal solution? J. Doc. 57, 284–295. http://dx.doi.org/10.1108/EUM0000000007103.

Lawton, J. H., 1999. Are there general laws in ecology? Oikos 84, 177–192.

Levin, S. A., 1992. The problem of pattern and scale in ecology. Ecology 73, 1943–1967.

Lin, C.-C., Porter, J. H., Lu, S.-S., 2008. A metadata-based framework for multilingual ecological information management. Taiwan J. For. Sci. 21, 1–6.

Mackenzie-Robb, L., 2010. Controlled vocabularies vs. full text indexing. www.vantaggio-learn.com/White%20papers/vocabularies%20vs.%20text%20indexing.pdf.

Madin, J. S., Bowers, S., Schildhauer, M. P., Jones, M. B., 2008. Advancing ecological research with ontologies. Trends Ecol. Evol. 23, 159–168.

Madin, J. S., Bowers, S., Schildhauer, M. P., Villa, F., 2007. An ontology for describing and synthesizing ecological observation data. Eco. Inform. 2, 279–296.

Mai, G. S., Wang, Y. H., Hsia, Y. Y., Lu, S. S., Lin, C. C., 2011. Linked open data of ecology (LODE): a new approach for ecological data sharing. Taiwan J. For. Sci. 26, 417–424.

Michener, W. K., Jones, W. B., 2012. Ecoinformatics: supporting ecology as a data-intensive science. Trends Ecol. Evol. 27, 85–93.

Miles, A., Bechhofer, S., 2009. SKOS simple knowledge organization system reference. www.w3.org/TR/skos-reference (accessed April 3, 2016).

Ohte, N., Nakaoka, M., Shibata, H., 2012. ILTER and JaLTER: their missions and linkage to database development in the Asia-Pacific region. In: Nakano, S., Yahara, T., Nakashizuka (Eds.), The Biodiversity Observation Network in the Asia-Pacific Region. Springer, Tokyo, pp. 206–216.

Porter, J., 2006. Improving data queries through use of a controlled vocabulary. LTER databits http://databits.lternet.edu/spring 2006/improving-data-queries-throughuse-controlled-vocabulary (accessed November 20, 2016).

Porter, J., 2010. A controlled vocabulary for LTER datasets. LTER databits. http://databits.lternet.edu/spring-2010/controlled-vocabulary-lter-datasets (accessed April 28, 2016).

Porter, J., Costa, D., 2006. Keywords and terms from the LTER Network – 2006. Long Term Ecological Research Network. http://dx.doi.org/10.6073/pasta/270a615ebb ecf90aebc72134a1bda355.

Rajbhandari, S., Keizer, J., 2012. The AGROVOC concept scheme – a walkthrough. J. Integr. Agric. 11, 694–699.

Schentz, H., Peterseil, J., Bertrand, N., 2013. EnvThes – interlinked thesaurus for long term ecological research, monitoring, and experiments. In: von Bernd, P., Fleischer, A. G., Gobel, J.,Wohlgemuth, V. (Eds.), EnviroInfo 2013: Environmental Informatics and Renewable Energies, 27th International Conference on Informatics for Environmental Protection – I and II. Shaker-Verlag, pp. 824–832.

Shahi, D., 2015. Apache Solr: an introduction. In: Apache Solr. Apress, pp. 1–9.

Tonn, W. M., Magnuson, J. J., Rask, M., Toivonen, J., 1990. Intercontinental comparison of small-lake fish assemblages: the balance between local and regional processes. Am. Nat. 136, 345–375.

van der Werf, D. C., Adamescu, M., Ayromlou, M., Bertrand, N., Borovec, J., Boussard, H., Cazacu, C., Van Daele, T., Datcu, S., Frenzel, M., Hammen, V., 2008. SERONTO, A Socio-Ecological Research and Observation Ontology: The Core Ontology. In: Weitzman, A., Belbin, L. (Eds.), Proceedings of TDWG, October 19–24, Freemantle, Australia, pp. 17–25.

Vanderbilt, K. L., Blankman, D., Guo, X., He, H., Lin, C. C., Lu, S. S., Ogawa, A., Ó Tuama, É., Schentz, H., Su, W., 2010. A multilingual metadata catalog for the ILTER: issues and approaches. Eco. Inform. 5, 187–193.

Vanderbilt, K. L., Lin, C.-C., Lu, S.-S., Kassim, A. R., He, H., Guo, X., San Gil, I., Blankman, C., Porter, J. H., 2015. Fostering ecological data sharing: collaborations in the International Long Term Ecological Research Network. Ecosphere 6, 1–18

Vihervaara, P., D'Amato, D., Forsius, M., Angelstam, P., Baessler, C., Balvanera, P., Boldgiv, B., Bourgeron, P., Dick, J., Kanka, R., Klotz, S., Maass, M., Melecis, V., Pet řík, P., Shibata, H., Tang, J., Thompson, J., Zacharias, S., 2013. Using long-term ecosystem service and biodiversity data to study the impacts and adaptation options in response to climate change: insights from the global ILTER sites network. Curr. Opin. Environ. Sustain. 5, 53–66.

Walls, R. L., Deck, J., Guralnick, R., Baskauf, S., Beaman, R., Blum, S., Bowers, S., Buttigieg, P. L., Davies, N., Endresen, D., Gandolfo, M. A., Hanner, R., Janning, A., Krishtalka, L., Matsunaga, A., Midford, P., Morrison, N., Ó Tuama, É., Schildhauer, M., Smith, B., Stucky, B. J., Thomer, A.,Wieczorek, J.,Whitacre, J.,Wooley, J., 2014. Semantics in support of biodiversity knowledge discovery: an introduction to the biological collections ontology and related ontologies. PLoS One 9 (3), e89606. http://dx.doi.org/10.1371/journal.pone.0089606.

Wood, D., Zaidman, M., Ruth, L., Hausenblas, M., 2014. Linked Data. Manning Publications, Shelter Island, NY.

Part II

Translating and exploring environmental genres

Literature, media and social promotion

7 Exploring the translation, diffusion, and reception of Under the Dome in the media

Meng Ji

7.1 Introduction

Environmental documentary represents a specialist genre which requires not only investigative and critical analyses of environmental events, facts and social phenomena, but also the development of a better, shared understanding among the public of the causal relations between human and industrial activities and their environmental impacts. Well-received environmental documentaries may significantly raise the social awareness of the importance and the public endorsement of environmental protection at the grassroot level; and can thus be used as an important environmental promotion instrument and tool. In China, investigative documentaries that focus on environmental health risks have been very limited. Under the Dome by Chai Jing (2015) represents a milestone in the development of the genre of environmental documentary in China, as the Chinese journalist has created a unique style of environmental reporting that is characterized by her personal and feminine concerns over the health impacts of air pollution in China. The significance of Chai's documentary is that she successfully developed a highly persuasive style of environmental reporting, and that her particular language and communication style has fostered a strong sense of unshirkable social responsibility among individual Chinese citizens.

Investigative environmental documentary represents an emerging literary genre in China. This particular type of environmental literature has gained social popularity amidst growing public concerns over the health impacts of major environmental events and disasters such as air, water, soil and ocean pollution. Different from current investigatory environmental reporting that focuses on the critique of lack of governmental accountability of severe environmental issues in China, the Chinese environmental documentary Under the Dome (2015) by the former CCTV (China Central Television) journalist Chai Jing has developed a personal and persuasive communication style that offloads the social responsibility of environmental protection back to the public or each individual contributor to and consumer of China's modern lifestyle. From food waste in restaurant to the overconsumption of fossil fuels in major Chinese cities, Chai expounded that the noxious air pollution in China is largely attributable to the widely-existent and persistent lack of public knowledge and social awareness of the environmental impacts on human health. Before an avalanche, not a single snowflake feels they are responsible – using metaphors and real-life examples collected around the country, the Chinese environmental documentary illustrated the social process in which ordinary people become both the culprits and victims of China's looming environmental crisis. The off-loading of environmental responsibility, which is more commonly seen in developed economies like Japan (see Hook, Lester, Ji, et al., 2017) has been rarely discussed in the Chinese media. The Chinese public has so far been portrayed mostly as the victims of environmental pollution with limited access to environmental health information. In Chai's documentary, the logic that economic development and the lack government accountability were the only reasons for China's environmental crisis was challenged, revisited. Within a short period of release, the film won the public sympathy, official endorsement and international recognition. With the wide praise and increasing popularity of Chai's documentary, a new environmental culture promoting shared social responsibility and individual accountability to the whole society was about to turn China's environmental movements on its head. It was found that after the wide dissemination of Chai's documentary, there has been a surge in Chinese media of expressions and terms such as green ethics, responsible consumerism, green culture vanguard, eat-up mission (avoid food waste), carbon chastity, and so on.

Environmental movements in China have since then developed new dimensions that are associated with people's behavioural and lifestyle change. The culmination of the recent environmental developments is the integration of citizens' environmental ethics and a common environmental culture – or the Community of Shared Future for Mankind (renlei mingyun gong tong ti) – into the modernization and social development agenda of China (Xi, 2017). While Chai's environmental documentary has been discussed, studied extensively for its success inside of China, there has been limited research of its translation, diffusion and reception overseas. The aim of this study is to provide an empirical linguistic analysis of the social reception of the translation of Under the Dome by Chai among English-speaking audiences, in an effort to gauge and contrast the social impact that the documentary has created internationally comparing to its domestic reception. The study of the social reception of the translation of Under the Dome involves the study of three types of media resources, that is, mainstream newspapers, news-based social media and multimedia-based social media such as public comments posted on YouTube. The study of a variety of media and social media instead of a particular type of media materials will enable a balanced, inclusive evaluation of the social reception of the Chinese environmental documentary among audiences with distinct language, cultural and education backgrounds.

7.2 Construction of the UTD corpus

In order to offer a balanced and inclusive evaluation of the wide social reception of the translation of Chai's Under the Dome among English-speaking audiences, a variety of media resources were collected to build a corpus (UTD corpus hereafter) that contains large amounts of original social media data commenting on the Chinese environmental documentary. Specifically, the UTD corpus constructed encompasses articles published between 2015 and 2018 on Chai's Under the Dome in major newspapers in the United States, the United Kingdom and Australia; public comments associated with these news (can be found beneath the news articles); and relevant social media data on popular multimedia platforms such as YouTube. The UTD corpus has incorporated three major types of public opinion data, that is, news articles, news-based social media data and multimedia-based social media data. The mainstream newspapers collected into the UTD corpus include the *Washington Post*, the *Wall Street Journal*, the *New York Times*, the *Financial Times*, *The Daily Mail*, NPR News, *The Guardian* (Australia), ABC News (Australia) and the *Sydney Morning Herald*. These are the major English newspapers which have reported on Chai's environmental documentary. The UTD corpus also contains large amounts of news-derived social media data associated with the coverage of Chai and her documentary in English mainstream media. Lastly, the UTD corpus has included all the public posts ascribed to the most-viewed 11 YouTube video sessions of the Chinese environmental documentary, all of which were provided with English subtitles to facilitate the understanding of the Chinese audio-visual materials. In linguistically-oriented translation studies, the development and exploration of language corpora offers an opportunity to discover the underlying patterns and textual structure of translations.

The central aim of this study is to analyse the social reception of the translation, dissemination of key environmental messages in Under the Dome among the target English audiences. In the context of this study, we adopt a broad definition of translation to include both digital news articles featuring abridged, adapted translated segments of the original Chinese documentary, and the audiovisual form of translation (or subtitle) available on popular multimedia platform such as YouTube (see Table 7.1). This allows the collection of relevant media and social media data from a wide range of sources, instead of using resources that reflect the views and opinions from specialist audiences only. This chapter makes an original contribution to the study of environmental translations by developing a corpus of social media data to investigate and configure the structure and key dimensions of the public opinions and attitudes towards the Chinese environmental documentary under study. The complete UTD corpus contains

Table 7.1 Structure of the media translation corpus of Under the Dome (UTD)

Media Communication Mode	Translation Mode	Associated Social Media Type
Newspapers	Translation of the original script	Newspapers-based public posts
Multimedia	Subtitles	Multimedia-based public posts

over 200,000 words from 29 different sources covering mainstream media such as newspapers, news-based social media and multimedia-based social media. The large media database makes useful preparations for the corpus-based analysis of the main dimensions of the social response and reception of Under the Dome among the English-speaking audiences. In order to explore the psychological properties of the country cohorts of media and social media data, the raw corpus data in the UTD is then subjected to the systematic linguistic annotation using popular social media data annotation tools, that is, the Linguistic Inquiry and Word Count (LIWC) annotation tool was used in this study. The next section will offer detailed explanations of the various annotation categories of LIWC, and the rationale behind its annotation scheme based on latest research in psychological linguistics, corpus linguistics and social media data mining.

7.3 Exploring sentiment and psychological properties of the social media corpus of Under the Dome

In corpus linguistic research, the annotation of language and textual corpora may be carried out at different levels. The commonly-used corpus annotation schemes include syntactic, morphological and semantic annotation. While these annotation programmes have proved to be efficient and reliable with formal written materials, their validity and effectiveness in the study of personal and informal written materials such as social media remain to be ascertained. The Linguistic Inquiry and Word Count (LIWC) annotation system has been widely used in psychological research since the 1990s. The latest version of the system runs on a large internal dictionary LIWC 2015, which contains words associated with psychologically-relevant categories such as analytical thinking, relative social status or clout, authenticity or honesty and emotional tones.

Apart from these abstract meta-linguistic variables, the LIWC also provides a breakdown of linguistic and textual features in fourteen different domains: language metrics, function words, grammatical categories, affect words, social words, cognitive processes, perpetual processes, biological processes, core drives and needs, time orientation, relativity, personal concerns, informal speech and punctuations. Within each of the fourteen textual and linguistic categories, a number of subcategories were highlighted to enable an integral and multi-dimensional analysis of the communicative properties of the text input. For the current study, the large variety of text and linguistic annotation categories and subcategories will enable the corpus-driven exploration of the psychological structure of the social media database; and help to probe and establish the association between the combination of various textual and linguistic features and the four predefined psychological scales of the LIWC.

7.4 Data analysis and discussion

After the automatic annotation of Under the Dome using the LIWC, the tagging system provides a breakdown of the distribution of different subcategories in each

Table 7.2 Subcategories of LIWC

	Subcategories (tagging labels)
Language metrics	words per sentence (WPS); words longer than 6 letters (sixltr); dictionary words (dic);
Function word (function)	total pronoun (pronoun); personal pronoun (ppron); 1st person singular (I); 1st person plural (we); 2nd person (you); 3rd person plural (they); impersonal pronoun (ipron); articles (article); prepositions (prep); auxiliary verbs (auxverb); common adverbs (adverb); conjunctions (conj); negation (negate); regular verbs (verbs); adjective (adj); comparative (compare); interrogatives (interrog); numbers (number); quantifiers (quant)
Grammar	regular verbs (verb); adjectives (adj); comparatives (compare); interrogatives (interrog); numbers (number); quantifiers (quant)
Affect words (affect)	positive emotion (posemo); negative emotion (negemo); anxiety (anx); anger (anger); sadness (sad)
Social words (social)	family (family); friends (friend); female referents (female); male referents (male)
Cognitive processes (cogproc)	insight (insight); cause (cause); discrepancies (discrep); tentativeness (tentat); certainty (certain); differentiation (differ)
Perpetual process (percept)	seeing (see); hearing (hear); feeling (feel)
Biological process (bio)	body (body); health/illness (health); sexuality (sexual); ingesting (ingest)
Core drives & needs (drives)	affiliation (affiliation); achievement (achieve); power (power); reward focus (reward); risk/prevention focus (risk)
Time orientation	past focus (focuspast); present focus (focuspresent); future focus (focusfuture)
Relativity (relativ)	motion (motion); space (space); time (time)
Personal concerns	work (work); leisure (leisure); home (home); money (money); religion (relig); death (death)
Informal speech (informal)	swear words (swear); netspeak (netspeak); assent (assent); nonfluencies (nonfl); fillers (filler)
Punctuations (allpunc)	periods (Period); commas (Comma); colons (Colon); semicolons (SemiC); question marks (QMark); exclamation marks (Exclam); dashes (Dash); quotation marks (Quote); apostrophes (Apostro); parentheses (Parenth); other punctuation (OtherP)

of the twenty-nine subfolders of the corpus. Table 7.3 shows a breakdown of the scores of the four psychological scales attributed to the media and multimedia data collected in the Under the Dome corpus. As described earlier, the four psychological scales represent composite measurements which have been developed from previous psychological and behaviour research. These psychological properties of media and social media materials are expressed and constructed through the combined use of the eight-three linguistic, textual and discourse features and devices (see Table 7.2). The scale of analytical thinking contains function word categories which measures the writing styles of different media reporters and social media contributors. Higher analytical thinking scores indicate more logical, formal and hierarchical thinking patterns and lower analytical thinking scores are suggestive of more narrative, personal communication patterns. Table 7.3 shows that for newspapers, the analytical thinking scores are consistently high ranging between 81.72 and 99.0 with an average score at 93.60. This is followed by newspapers-derived social media, which has an overall lower mean score at 76.76. Public posts collected from popular multimedia platforms such as YouTube has the lowest average score of 70.48. Important variation was detected in the analytical thinking scores of multimedia data which ranged between 42.61 and 92.84.

In the 2015 version of LIWC, the psychological scale of clout refers to the relative social status among people and the self-confidence that people express through the use of word categories such as pronouns. The psychometric scale of Clout is distinct from the word category of Power under Core Drives and Needs (see Table 7.2). Higher clout scores may serve as an indication of greater confidence, interaction among the social media users or a more personal and direct reporting style of newspapers. Previous studies show that low status members of Internet message boards tend to use a higher frequency of the first single pronoun 'I' than a lower frequency of the second single pronoun 'You' compared to higher status members (Dino, Reysen & Branscomb, 2008). The use of 'I' has also been found to be negatively associated with other perceptions of dominance (Berry, Pennebaker, Mueller & Hiller, 1997). The average score of Clout for newspapers is 70.72 which is higher than the Clout scores for the two types of social media under study: 63.83 (multimedia-derive social media posts) and 62.00 (news-based social media posts). An example which illustrates the distinct communicative styles of newspapers and social media is found between *The Daily Mail* (UK) and YouTube 11 (ranked by the total number of views, with YouTube 1 as the top-viewed version of Under the Dome with English subtitles). The distribution of personal pronouns in The Daily Mail and YouTube11 displays contrastive patterns: 'I' (0.26% Daily Mail and 2.63% YouTube 11); 'We' (0.26% Daily Mail and 0.00% YouTube 11); 'You' (0.13% Daily Mail; 2.63% YouTube 11); 'She/He' (4.90% Daily Mail; 0.00% YouTube 11); 'It' (3.23% Daily Mail; 5.26% YouTube 11). The contrastive patterns suggest that viewers of the multimedia-based translation of the Chinese environmental documentary are more likely to express their comments from their personal experiences and use that experience to interact with the fellow viewers. By contrast, major newspapers tend to frame and deliver the translated messages of the original documentary from distance as evidenced in the higher use of third singular pronouns she/he.

Similar to Analytical and Clout, the psychometric scale of Authenticity represents a composite scale that combines a few linguistic and textual sub-categories from Table 7.1 based on previous research (Newman, et al. 2003). Specifically, this scale encompasses first-person singular pronoun 'I/me/mine', third-person pronouns 'he/she/they/their/them', negative emotion words (such as regrettable, paranoia, hazard, cruel, helpless, serious, pathetic, ashamed, etc.), differentiation words (such as different, than, against, but, instead, etc.), and motion verbs (such as put, grow, explore, push, deliver, etc.). The psychometric scale of Authenticity was originally developed for the purpose of detecting and predicting deception in people when discussing controversial or private issues such as abortions or relationships (Newman, et al. 2003). A higher Authenticity score is associated with greater truthfulness and honesty. Table 7.2 shows that newspapers have the average Authenticity score of 30.40, which is largely comparable to the average Authenticity scores of news-derived social media (33.70) and multimedia-based posts (29.21). The relevance of this psychometric scale to the current study is not as immediate as the other three scales. This can be hinted at the similar Authenticity scores of the three types of media under comparison, as this scale does not differentiate them well.

The last psychometric scale is known as Emotional Tone, which includes both negative (lower than 50) and positive emotions (higher than 50). The average tones for newspapers, multimedia-based public posts and news-based social media are 34.18, 79.54 and 28.89, respectively. High-frequency positive emotional words in multimedia-based public posts are "important, brave, true, care, create, improve, agree, inspiring, surprising, amazing, accept, respected, bravery, impressed, hope, freed, incentives, good, liberty, support, strong, appreciate, improvement, challenging, wisdom, opportunity, promise, better, courage, courageous, active, useful, original, benefit, healthy, encourage, honest, libertarians, wellbeing, optimal, success, energies, readiness, share, warming, best."

The lowest average emotional tone was found with newspaper-based public posts and expressions associated with negative emotions in this type of social media include "weak, denial, pity, greed, alarming, problem, strange, sad, desperate, disaster, incompetent, destroy, damage, destruction, warning, worst, offender, ashamed, problem, frustration, serious, attacks, alone, complaining, awful, pathetic, sinister, low, embarrassment, worrying, gross, bad, ignore, shock, criticise, worsening, unfair, overwhelming, poisoned, crazy, poor, depressing, ridiculous, abandoned, threating, shamefully, difficult, nasty, suffering, struggle, degradation, fight, isolated, punishing, hazardous, selfish, aggressive, deceiving, ignorant, fooled" and so on. It became clear from the two lists of positive and negative emotional expressions that the focus of the discussions among multimedia users and viewers of the subtitled version of Under the Dome was on the courage of the reporter and the inspiration of the documentary.

This group of viewers or consumers of the audio-visual version of the translation of Under the Dome seems to have developed a shared consensus of the opportunities promised by this film to raise social and public awareness around environmental problems and their possible solutions, that is, the active and proactive contributions from individual citizens to clean up the air pollution in China. This positive reaction identified in the multimedia-based public posts echo the positive social impact produced by the original Chinese environmental documentary among its Chinese audiences. The list of positive emotional expressions from the English audiences and viewers explicitly points to the solidarity and endorsement of Chai's critical analysis of China's modern lifestyle, its environmental impact and the sharp increase in cancers and critical health conditions among vulnerable social groups such as children, elderly people and people from lower socio-economic backgrounds. The somewhat low emotional tone of readers of newspapers reporting Chai's documentary may well be influenced by the low emotional tone of the newspapers containing translated messages and segments from the Chinese documentary.

This group of readers rely essentially on newspapers or mainstream media as their main sources of information, instead of full translations of the Chinese original. It is also useful to note that while in Australia, the tone of readers and users of news-based social media is significantly higher than the tone of newspapers reporters; the reversed pattern is consistent with the British and American sub data sets. In other words, the linguistic analysis suggests that while the Australian news readers appear to be more sympathetic and supportive of Chai's environmental documentary and its wider social impact, the British and American social media users and readers of newspapers reporting on Under the Dome are more critical and focused on the environmental and social events reported in the documentary. This clear divergence implies two distinct approaches to and modes of the social reception of the film characterized by the emotional interaction with the Chinese environmental reporter leading the narration of the documentary; and the criticism of the socio-economic development, and the environmental and health impacts of people's behavioural and lifestyle change in China.

The corpus analysis shown earlier was chiefly based on predefined psychometric scales which have been developed in previous psychological and behaviour research. While some of the scales such as analytical thinking and emotional tones were more useful and relevant for the current study, the applicability and discriminatory power of some scales such as Authenticity and Clout has proved to be limited. This suggests that the existing analytical scales of LIWC do not apply readily in the study of the reception of the two types of translations of Chai's environmental documentary, that is, new articles containing translated segments such as script of the narrator of the original Chinese documentary and the subtitled audio-visual version of Under the Dome. In order to better understand the differences among these two types of translations that have been leveraged in the media reporting of air pollution in China, the following section will use the data-driven approach to attempt to exploratory and analyse the internal structure and key dimensions of the translation, social diffusion and reception of Chai's Under the Dome among the target English audiences.

7.5 Exploration of key dimensions of the social reception of UTD (PCA)

The use of the LIWC system in of the annotation of Under the Dome media corpus generated to a very large number of quantifier linguistic and textual features, which amounts to 83 categories in total (see Table 7.2). The analysis and comparison of the media and social media data sets in the UTD corpus by

Table 7.3 Psychological scales of LIWC: Analytical, Clout, Authentic and Emotional Tone

	Country	Media Type	Analytical	Clout	Authentic	Tone
ABC News	Australia	Newspapers	91.15	74.42	20.72	32.60
Sydney Morning Herald	Australia	Newspapers	99.00	71.40	33.40	15.42
The Guardian	Australia	Newspapers	92.38	52.25	52.64	20.64
NPR	US	Newspapers	81.72	72.98	27.32	43.98
The Daily Mail	UK	Newspapers	95.99	88.98	21.46	39.62
Financial Times	UK	Newspapers	98.76	70.38	49.53	61.05
New York Times	US	Newspapers	92.91	66.24	25.52	27.01
Wall Street Journal	US	Newspapers	94.72	71.78	16.01	47.39
Washington Post	US	Newspapers	95.80	68.04	27.04	19.90
Youtube1	International	Multimedia	72.09	63.10	24.58	60.33
Youtube2	International	Multimedia	68.47	67.93	16.37	69.03
Youtube3	International	Multimedia	67.02	46.18	37.62	55.81
Youtube4	International	Multimedia	63.78	70.51	53.71	87.42
Youtube5	International	Multimedia	92.84	64.35	5.55	99.00
Youtube6	International	Multimedia	42.61	85.00	16.43	47.90
Youtube7	International	Multimedia	54.18	59.55	35.25	84.42
Youtube8	International	Multimedia	76.65	73.73	13.15	95.81
Youtube9	International	Multimedia	79.40	87.67	14.27	99.00
Youtube10	International	Multimedia	82.00	50.00	13.65	78.60
Youtube11	International	Multimedia	76.19	29.92	90.72	97.58
ABC News	Australia	News-based SM	59.74	69.50	31.08	35.34
Sydney Morning Herald	Australia	News-based SM	94.22	66.72	41.78	25.77
The Guardian	Australia	News-based SM	68.65	46.49	51.65	32.46
NPR	US	News-based SM	60.01	55.46	44.46	18.28
The Daily Mail	UK	News-based SM	70.84	72.12	25.23	22.68
Financial Times	UK	News-based SM	79.88	66.47	23.85	40.31
New York Times	US	News-based SM	84.00	61.60	30.49	34.81
Wall Street Journal	US	News-based SM	85.00	63.24	20.72	49.37
Washington Post	US	News-based SM	88.54	56.30	34.05	7.47

each of the 83 linguistic subcategories would be a lengthy and counterproductive process. Some of the linguistic and textual features are internally correlated and can be extracted from the pool of annotated corpus categories and integrated into a general analytical scale or dimension. For example, the corpus analysis that follows shows that verbs, function words, auxiliary verbs and adverbs were found to be used more frequently in the media and social media data sets of the UTD corpus which have a higher analytical thinking score, for instance, major newspapers.

These linguistic and textual features were extracted and grouped into a single analytical dimension in the statistical exploration of the media translation corpus. To streamline the grouping and classification of linguistic and textual features, principal component analysis (PCA hereafter) was used. PCA has assisted in the configuration of the key dimensions of the media data of the UTD corpus reflecting the social reception of the film among the target English-speaking audiences and readerships. Before using PCA, the raw data sets were transformed and standardized into their corresponding z-scores to deal with the issue of missing value for some of the LIWC word categories.

Table 7.4 shows the result of PCA, especially the internal structure of the streamed model which classifies the original 83 linguistic and textual features (see Table 7.2) into five large components shown here. The contribution of each of the five components to the new analytical model is indicated by the total amount of variances explained by that component. For example, the first component explains as much as one quarter (23.397%) of the entire data sets. It therefore represents the largest and most powerful analytical component of the quint-component model. This is followed by the second, third and fourth components, with each explaining around one fifth of the total variance of the UDT corpus. The fifth component proved to be the smallest contributor to the model accounting for one-tenth of the total variations in the media data sets. As a result, the total model can explain as much as over 80% of the total variations in the UTD corpus.

Table 7.5 shows the linguistic and textual features attributed to each of the five components of the streamlined model. The first component includes six grammatical and semantic categories which are verbs (0.946), function words (0.895), auxiliary verbs (0.891), adverb (0.797), cognitive processes (0.735) and certain (0.669). Numbers in parentheses are the component loading scores ranging between -1 and 1. The larger the positive loading score, the more the contribution a textual/linguistic feature makes to an analytical component. For example, the linguistic categories of verbs, function words, auxiliary verb (such as may, could, would, being, does, wouldn't, does, doesn't, doing, isn't, be, becomes, will, have/has been, won't, can, do, done, has, must, aren't, did, should, might, I'll, having, wasn't, etc.), adverbs, cognitive processes (such as possibly, want, allow, recognize, adjust, become, changes, choices, learn, consequences, mistakes, seem, appear, consider, hope, wish, prove, appreciate, accept, suggest, hint, cause, create, think, solve, guess, feel, know, need, imagine, remember, recall,

Table 7.4 Principal component analysis of the Under the Dome media corpus

Total Variance Explained

Component	Initial Eigenvalues			Extraction Sums of Squared Loadings			Rotation Sums of Squared Loadings		
	Total	% of Variance	Cumulative %	Total	% of Variance	Cumulative %	Total	% of Variance	Cumulative %
1	7.102	28.408	28.408	7.102	28.408	28.408	5.849	23.397	23.397
2	4.675	18.700	47.108	4.675	18.700	47.108	4.051	16.205	39.602
3	4.030	16.118	63.227	4.030	16.118	63.227	3.813	15.252	54.854
4	2.638	10.553	73.779	2.638	10.553	73.779	3.777	15.108	69.961
5	1.788	7.150	80.929	1.788	7.150	80.929	2.742	10.968	80.929

Table 7.5 Linguistic/textual features of the five PCA components

Components/ Media Language Style	Rotated Component Matrix[a]					
	Linguistic/Textual Features	Component				
		1	*2*	*3*	*4*	*5*
Reflective comments	Verb	0.946				
	Function words	0.895				
	Auxiliary verbs	0.891				
	Adverb	0.797				
	Power	−0.788				
	Words of six or more letters	−0.757				
	Cognitive Processes	.735				
	Certain	.669				
Informal discussion of economic factors of environmental problems	Money		.881			
	Non fluencies		.868			
	Informal speech		.833			
	Assent		.822			
	Anxiety		.731			
Emotion	Exclamation Mark			.889		
	Quantifiers			.864		
	Second personal pronoun 'You'			.756		
	Adjectives			.728		
Gender/family relevance of environmental health issues	Female referents				.917	
	Family				.889	
	Third personal pronouns 'She/he'				.843	
	Social words				.816	
Personal sensory experiences	See					.911
	Perpetual process					.908
	Leisure					.695

Extraction Method: Principal Component Analysis.
[a]Rotation Method: Varimax with Kaiser Normalization.

pretend, suppose, try, understand, realize, question, notice, expect, believe, trust, suspect, sense, reason, distract, find out, wonder, doubt, decide, analyse, etc.) and expressions of certainty (undeniably, clearly, complete, proven, obvious, entirely, absolutely, precisely, true, must, definitely, exactly, always, never, ever, truth, sure, fact, etc.) have large positive loadings on the first dimension. It should also be noted that the two linguistic categories of Power and Words of Six or More Letters have large negative loadings on this dimension. These two categories tend to be associated with formal and analytical writing style of political news. Their negative loadings suggest that the first component is better described and summarized as reflective or cognitive thinking rather than analytical and complex thinking characteristic of political news.

Examples:

1. *Been* (auxiliary verb) *there* (adverb) (China) *seen* (verbs) that. Breathed that, *had* (auxiliary verb) sore eyes, *got out* (verbs) as soon as I *could* (auxiliary verb). Pity, it's a *very* (adverb) historic place. A *need* (cognitive process) for worldwide, fundamental change! Has *never* (certainty expressions) been *more* (adverb) *needed* (cognitive process)! (public posts on the ABC News, Australia)

2. I *might* (auxiliary verb) *note* (cognitive process) that the London smog which are *so often irrelevantly* (adverbs) *cited* (verbs) as part of tu quoque *arguments* (cognitive process) were *done* (auxiliary verbs) away with without the UK *reverting* (verbs) to *living* (verbs) in pre-capitalist poverty, and that the Chinese government has *spent* (verbs) decades with the technology in hand to *solve* (cognitive process) or at least *mightily* (adverbs) *mitigate* (verbs) its *problems* (cognitive process) without *using* (verbs) it. I *might* (auxiliary verbs) *note* (cognitive process) that even the Chinese government *claims* (verbs) it's now *going to do* (auxiliary verbs) something, and *probably* (certainty expressions) *doesn't have* (auxiliary verbs) *impoverishing* (verb) the country on its list, but rather smoke-stack scrubbers and similar. (Public posts on Sydney Morning Herald, Australia)

3. *Visit* (verbs) the Far East and pollution *is* (auxiliary verbs) *so* (adverb) *obvious* (certainty expressions) in many of its countries. You *don't* (auxiliary verbs) *need* (cognate process) scientific instruments to *detect* (cognitive process) it. This lady *brings* (verbs) a ray of *hope* (cognitive process) through the mist and we *hope* (cognitive process) the message *will* (cognitive process) *get through* (verbs). (Public posts on The Daily Mail, UK)

4. Oh, you *shouldn't* (auxiliary verbs) *have been* (auxiliary verbs) *so* (adverbs) *shocked* (cognitive process). We are *talking* (verbs) about the *possible* (certainty expressions) involvement of a country whose involvement in another country's humanity affair *could* (auxiliary verbs) *entail* (verbs) the mass injection of syphilis to the citizens of that victimized country. Your *apparent* (certainty expressions) innocence *is* (auxiliary verbs) *so* (adverbs) touching. (Public posts on Financial Times, UK)

5. I *watched* (verbs) it (the documentary) *all* (adverbs). Although the pace was a little fast and the English cc needed further editing, it was *intelligently* (adverbs) *written and produced* (verbs) and *may be* (auxiliary verbs) the beginning of a movement *urgently* (verbs) *needed* (cognitive process). The *creator* (cognitive process) *provided* (verbs) some easy and not so easy fixes, and *may have* (auxiliary verbs) made the CCP wobbly when she *gave out* (verbs) the toll free phone number for *complaints* (cognitive process). Despite the CCP *response* (cognitive process) of *blocking* (verbs) its distribution, "Under the Dome" will get around and begin spurring *some long* (adverbs) *needed* (cognitive process) action. Not in time to save the millions who *will* (auxiliary verbs) *perish* (verbs) over the next 20 years, but better late than *never* (certainty expressions). (Public posts on New York Times, US)

6. *Unbelievably* (certainty expressions) *well* (adverbs) *researched* (verbs), *well* (adverbs) *presented* (verbs) and relevant. *Probably* (certainty expressions) the *very* (adverb) best documentary I *have* (auxiliary verbs) *ever* (certainty expressions) seen (verbs). *Having* (auxiliary verbs) *spent* (verbs) the past year *wearing* (verbs) a PM2.5 mask while *cycling* (verbs) with work in Shanghai *pretty much* (adverbs) every day for the past year, and *having* (auxiliary verbs) *talked* (verbs) smog every time I *found* (cognitive process) myself with a journalist, an economist, a chemical, energy or automotive exec, or government official (basically every day) *I'm* (auxiliary verbs) *stunned* (verbs) at how *well* (adverbs) Chai Ling *covers* (verbs) this. She *brings* (verbs) all the pieces into context of regulatory and political reality, as well as quality of life context. She *asks* (cognitive process) the smart questions and *drives* (verbs) the answers to quantitative and political clarity. Second hard hitting, intelligent, important yet not angry documentary. *I've* (auxiliary verbs) *watched* (verbs) in two days. Plastic Paradise (about the Great Pacific Plastic Patch) was the other. Both *made* (verbs) by women, who *happen* (verbs) *to be* (auxiliary verbs) Chinese/Chinese American. Women *can* (auxiliary verbs) *just* (adverbs) *do* (verbs) *more* (adverbs). (Public posts on YouTube)

The second component includes five linguistic and textual features which are large contributors to this analytical dimension: money (0.881), non-fluencies (0.868), informal speech (0. 833), assent (0.822) and anxiety (0.731). The category of money includes expressions such as dollar, economy, rich, cheaper, consumers, profit, purchasing, buys, cheap, wages, pay, spend, accountable, businesses, debt, tax, expenses, price, savings, yuan (Chinese RMB), poverty, and so on. The linguistic category of non-fluencies comprises expressions such as oh, well, ugh, sigh, er, em and so on. Typical expressions in informal speech are dm, agree, damn, crap, anyway, cc, apps, shit, ok, hell, wanna, awesome, haha, gonna, cool, soo, and so on. The five linguistic and textual features with heave loadings on the second dimension suggest a common analytical dimension which is characterized by an informal communication style and a focus of discussions and debates of the relations between economic development and China's environmental crisis. This second component may be described and summarized as the informal discussions of the socio-economic stressors to the growing environmental problems in China.

Examples:

7. It's absolutely astonishing to realize, there's a whole generation of Chinese that have never seen a blue sky! Collateral damage for our *consumerism* (money) and *greed* (money), on a road paved by *Capitalism* (money) and driven by Communism?? Sad, sad state of affairs. No really, our dictator said "coal was good for humanity"…so it's *ok*! (informal speech/assent) … *um* (non-fluencies)… (Public posts on The ABC News, Australia)
8. There are many selfless journalists risking their very lives and under great *pressure* (anxiety expressions) from the start. If you're going to praise and

look up to someone then choose them. *Hell* (informal speech), choose some of the more well known Chinese activists. There are several facts you can count on for sure though, 1. Pollution will kill off many of the old and young, both of which are issues themselves (1 child policy). 2. Pollution will linger and devastate everything, *I mean* (informal speech) everything; 3. *Rich* (money) Chinese will move and *buy* (money) properties in other countries. (Public posts on YouTube 1)

9. "Last Sunday, the new environment minister, Chen Jing, compared the video to Rachel Carson's "Silent Spring" in an inaugural news conference with Chinese reporters in Beijing." *Oh boy* (informal speech)...he is exactly *right.* (assent) This is a nuclear bomb of an expose in China. Especially couched as it is in family and such. It's little wonder the CCP was *worried* (anxiety expressions) and slammed the breaks on its viewing... The government, with its self-interest tied to a rogue *Capitalist* (money expressions) bed, will not change things in and of itself. (Public posts on YouTube 2)

The third component contains four linguistic and textual features which are exclamation mark (0889), quantifiers (0.864), second personal pronoun 'You' (0.756) and adjectives (0.728). Typical expressions in the quantifier category include all, every, entirely, many, multiple, less, part, least, more, some, bit, any, series, most, average, much, each, majority, single, absolutely, somewhat, none, lot, another, few, amount, ton, entire, single, remaining, amounts, either, piece, and so on. The combined use or co-occurrence of words and expressions in these four linguistic expressions suggests an outburst of strong personal emotions. For example, in the social media sub data set of the UTD corpus, a number of comments were found which provide useful illustration of this third analytical dimension:

Examples:

10. If you're born there, *you* (second personal pronoun) have *little* (adjective) choice*!* (Exclamation mark) (public posts on ABC News, Australia)
11. I am sorry but *you* (second personal pronoun) can't go to China. Mum says No*!* (Exclamation mark) It will not be *good* (adjective) for *your* (second personal pronoun) lungs*!* (Exclamation mark) (public posts on ABC News, Australia)
12. Under communism, *millions* (quantifiers) starved in near silence, too. *Your* (second personal pronoun) political narrative is showing DM*!* (Exclamation mark) "China = Air Pollution." (public posts on The Daily Mail, UK)
13. Yet *you* (second personal pronoun) *lot* (quantifiers) think mankind isn't affecting global *warming* (adjective) with *hundreds of millions* ((quantifiers)) of cars! Pshhh... (public posts on The Daily Mail, UK)
14. *You* can't hide *your* face in *your* hands *forever* China*!* (Exclamation mark) At *some* (quantifiers) point *you* have to raise *your* head and tear down the wall of *your* own making and face the music. In *your* race to modernize, *you*

have contracted the *same* (quantifiers) ills that the United States has been recovering from in its Industriousness. There are solutions, but they grow costly the *long* (adjectives) *you* wait. And the *longer* (adjectives) *you* wait, the *more* (quantifiers) *tenuous* (adjectives) *your* hold on power becomes*!* (Exclamation mark) (public posts on New York Times)

15. Share *your* judgment about that *outstanding* (adjective) documentary *absolutely* (quantifier)*!* (Exclamation mark) (YouTube public posts)

16. I hope *more n more* (quantifier) *ordinary* (adjective) people would be *willing* (adjective) to take *practical* (adjectives) actions *even* (quantifier) is *only* (quantifier) for *your next* (adjective) generation*!* (Exclamation mark) (YouTube posts)

The **fourth dimension** of the new analytical instrument contains four sub-word categories: female referents; family; third personal pronouns 'she/he' and social words. Chai's Under the Dome is regarded widely as a milestone of the development of ecofeminism in China. Saddened by the discovery of a benign tumour in her foetus, which Chai attributed to major atmospheric pollutants, the Chinese journalist, sole investigator and narrator of the documentary was driven and passionate to build a shared social understanding of individual citizens' responsibility to alleviate China's looming environmental crisis. The film features extensive discussions of the health impact of environmental pollutions on children from the perspective of a mother. The following script was taken from 21:11 minute of the film:

> Every day I wake up, first thing I do is look at the Air Quality Index app on my phone. Use it to arrange my day. I wear my mask shopping, buying groceries, meeting with friends. I use tape to cover every window frame. When I take my kid out to get vaccines, I get scared when she so much as giggles for fear she's breathing in more pollution. Honestly, I am not afraid of dying. I just don't want to live this way anymore. So whenever someone asks me, "Why are you doing all of this anyway?" I tell them this is personal beef between me and the haze. I want to know where it's coming from. I want to get to the bottom of this.

In the media and social media gathered to study the social reception of the Chinese documentary, Chai's personal touch of the topic provided another important focus of discussions among the target English-speaking audiences. The media database constructed has enabled the retrieval of original discussion segments which can be readily classified into and illustrated by this analytical component:

Examples:

17. I'm stunned at how well Chai Ling covers this. *She* (third personal pronoun/ female referent) brings all the pieces into context of regulatory and political reality, as well as quality of life context. *She* (third personal pronoun/female

referent) asks the smart questions and drives the answers to quantitative and political clarity. I've watched in two days. Plastic Paradise (about the Great Pacific Plastic Patch) was the other. Both made by *women* (female referent), who happen to be Chinese/Chinese American. *Women* (female referent) can just do more. (Public posts on YouTube)

18. *We* (social words) were fortunate. The day *we* (social words) went to the Great Wall, was beautiful, sunny and clear. The next day, when *we* (social words) woke up at our hotel in Beijing, the yellow pea soup of pollution had rolled in and stayed with *us* (social words) from Beijing through *our* (social words) trip to Xian. It reminded me of the pictures that *you* (social words) used to see of the smog in California. I was grateful that *we* (social words) were only staying for a short time in that part of China. China must do something about its pollution. Not only is it shortening lives but it is already having a damaging impact on the *unborn* (family expressions). The rate of *birth* (family expressions) defects in *babies* (family expressions) born in China is much higher than would be expected in a *population* (social words) not exposed to the soup of chemicals that *mothers* (family expressions) to be are exposed to in China. (Public posts on New York Time)

19. Air pollution has become a focus of public discourse in China in the past few years as thick smog has blanketed large swathes of the country. Chai says in the documentary *she* (third person pronoun) became afraid of pollution for the first time after *her* (third person pronoun) *baby daughter* (family expressions) was given an operation soon after *birth* (family expressions) to remove a benign tumour. *She* (third person pronoun) displays pollutant charts and footage of doctors removing blackened lymph nodes from the lungs of a *patient* (family expressions) who didn't smoke but was diagnosed with early-stage lung cancer.

Lastly, the **fifth component** contains three word categories that indicate the levels of personal involvement: see; perpetual processes and leisure. The word category of see contains expressions such as shows, viewership, views, blind, watch, see, look, visible, eye, sight, and so on. Typical expressions in the perpetual processes of the LIWC annotation system include say, look, disappear, smell, sweet, beautiful, looking, pic, show, taste, blue, photos, view, weight, blindly, sound, image, watch, videos, listen, hear, sense, experience, touch, vision, feel, sight, sing, touch, burning, yellow, silent, heavy, hurting, harsh, painful, rotten, visible, voices, clear, blurred, voice, clear, cold, warm, grab, black, dark, grey, odours, strong, scratch, speak, smooth, colour, green, fresh, loudly, speech, light, shadow, musical, deaf, hard, soft, thick, and so on. Expressions that have been classified under the leisure subcategory include movies, parties, dream, YouTube, videos, twitter, Facebook, TV, entertaining, documentary, books, travels, weekends, holidays, bike, swim, theatres, play, poetry, artists, games, comedy, teams, run, play, shopping, and so on. The fifth component thus underscores another main type of discussions reflecting the social reception of Under the Dome among the target audience and readership. This type of discussions is characterized by the discussants' personal and sensory experiences of the environmental events and phenomena reported in the Chinese documentary.

Examples:

20. I *watched* (see category) and *listened* (perpetual process) to the first 15 minutes of the *documentary* (leisure) and found (perpetual process) a combination of good science and very *strong* (perpetual process) *emotion* (perpetual process). *Hard* (perpetual process) to *imagine* (perpetual process) that the growing educated Chinese middle class who mostly live in these polluted cities will not be strongly moved (perpetual process) by the *video* (leisure). This supports the China-US agreement on climate and indicates that the Chinese leaders are supporting a strong shift away from polluting fossil fuels. (Public posts on The Guardian, Australia)

21. I'd *suspect* (perpetual process) the government never *expected* (perpetual process) it to go so viral! I've just *noticed* (perpetual process) there might be some very basic censorship as well; when you *search* (perpetual process) for it (in Chinese) on *youku* (leisure), nothing comes up but the *TV series* (leisure), even when you add Chai Jing's name! And a flood of more official, less *sensitive* (perpetual process) *videos* (leisure) about pollution have been released to fill the vacuum! (Public posts on The ABC News, Australia)

22. I *love* (perpetual process) the internet age, because even as governments fight I have *met* (perpetual process) so many people from across the world on *video games* (leisure) and *social networks* (leisure), I have *met* (perpetual process) people in almost every country, even in China, Iran and Russia. I really hope now that we can bridge the gap and *communicate* (perpetual process) with people who are across the world from us just by *hitting* (perpetual process) the "translate" button we can end all this bullshit squabbling between governments that the people don't want. We all want to fix the planet and help get rid of pollution before it's too late. (Posts on YouTube)

23. *Assume* (perpetual process) the massive *tencent* (leisure) *video* (leisure) *viewership* (see word category) of chai jing's pollution *documentary* (leisure) due to fact it is racing through *wechat* (leisure). Nearly 40m *views* (see word category) alone on *tencent* (leisure) *video* (leisure) for chai Jing's pollution *documentary* (leisure). (Public posts on Sydney Morning Herald, Australia)

24. 15 minutes in chai jing *documentary* (leisure) on air pollution. The *animation* (leisure) of how it attacks your *body* (perpetual process)…wow… who won't be *enraged* (perpetual process) *watching* (see word category) this? (Public posts on The Daily Mail, UK)

25. One needs to *wonder* (perpetual process) if Obama (or even an adviser to him) *saw* (see word category) this *movie* (leisure). I'm sure that Al Gore has *looked* (see word category) at it (LOL). Any sane person who *watches* (see word category) it cannot but be *impressed* (perpetual process) by the environmental disaster that is China. The NASA *photo* (perpetual process) proves that if we don't insist that China cleans up its act NOW, they and we will

suffer (perpetual process). China worse than us, but we'll get the fall-out. LA has worked to clean up its smog but Chinese pollution could bring it back. Maybe *Hollywood* (leisure) can promote this *film* (leisure). (Posts on Wall Street Journal)

The examples given here illustrate five large types of social responses and reactions to the social diffusion of Chai's environmental documentary Under the Dome among English-speaking audiences. The significance of these five types of media languages consists that they represent as much as over 80% of the media corpus constructed which contains over 200,000 words of media and social media data. The use of principal component analysis extracted the patterns and structures of five types of media languages. The next section will illustrate how these five types of media messages can separate social responses to Under the Dome, stimulated by news-based and multimedia translations.

7.6 Discussions of findings

Figure 7.1 shows the comparison of the social responses to Under the Dome in newspapers, news-based public posts and multimedia- based public posts which reflect the understanding and reception of Chai's Under the Dome. The criterion of comparison is the first component or media language communication style of the five-dimensional framework developed from using principal component ana-lysis (see Table 7.2). As described earlier, the first component was described and summarized as reflective or cognitive rather than critical or analytical thinking typical of political news. It became evident on this graph that most of the public comments collected from news-based social media and multimedia have posi-tive scores and cluster on the right side of the scale. By contrast, most of the news collected in the Under the Dome database have negative scores on the first dimension and are scattered on the left half of the graph. This finding suggests that the reflective and cognitive thinking and commenting style is shared among news readers and viewers of multimedia contents that feature the Chinese envir-onmental documentary with English subtitles.

This type of social media response is underlined by the higher frequency of use of linguistic categories such as adverbs, verbs, auxiliary verbs, function words, words indicting cognitive processes and expressions of certainty (see Examples 7.1– 7.6). News on Chai's Under the Dome however chiefly adopt more critical and analytical discussion style which feature the use of power-related expressions such as empowered, emasculated, strengthened, weakened; and words of six or more letters such as government, enormous, supervision, sensation, atmospheric, benign, responsibility, triggered, criticism, awareness, transportation, discontent, acknowledged, propaganda, sensitivity, consultation, authorities, administrative, parliamentary, phenomenon, solutions, cyberspace, impassioned, neutered, lucra-tive and so on which are abundant in the news data sets collected in the corpus.

Figure 7.2 shows the comparison of the news reporting and social media recep tion of Chai's documentary in Australia, the United States, the United Kingdom and

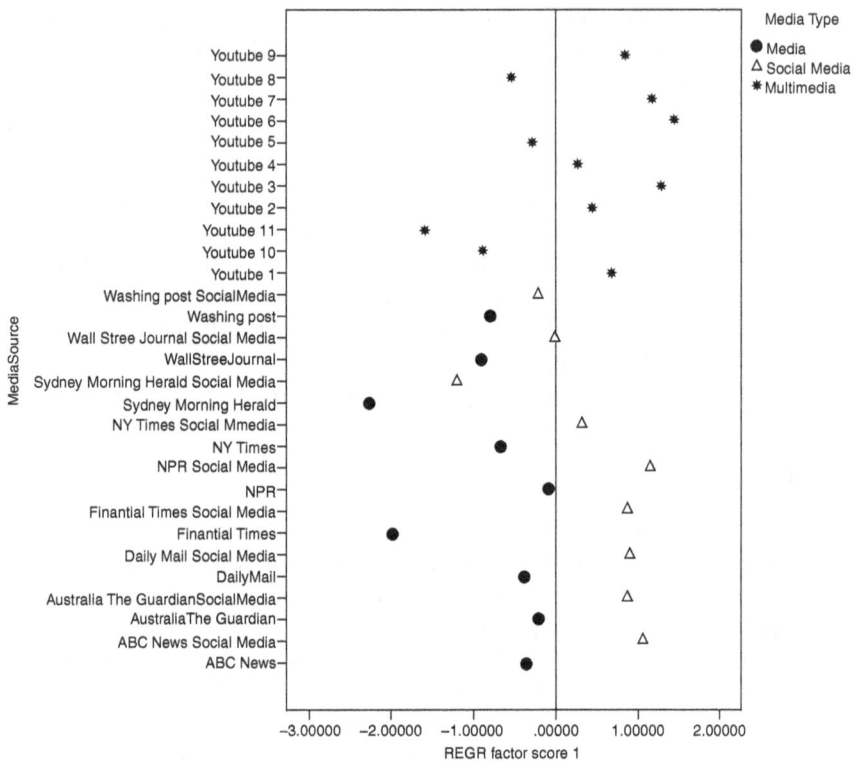

Figure 7.1 PCA Dimension 1: cognitive process.

internationally. The criterion used is the second component of the five-dimension analytical model, that is, informal discussion of the economic stressors of China's environmental problems. Higher positive scores on this scale indicate a higher frequency of use of the five component linguistic and textual categories which are expressions and words indicating money, non-fluencies, informal speech, assent and anxiety. By contrast, large negative scores on this scale indicate a more formal and less emotionally-involved communicative style. Illustration of the former type of public posts on social media is given in Examples 7.7–7.9. The (news-based) Social Media category in Figure 7.2 shows that readers of news in the United States (represented by a cross mark on the graph) (data collected from the *Washington Post,* the *New York Times,* the *Wall Street Journal* and NPR news) tend to adopt a more informal language style and seem to be more concerned with the economic factors of China's looming environmental crisis. Readers of the *Washington Post* have the highest positive score on this dimension: 0.72 indicating the higher use of informal language. This is followed by readers of the *Wall Street Journal* with an informal speech score of 0.40. The least informal language was found with readers posting on the *New York Times* who have the smallest informal speech score of 0.00083.

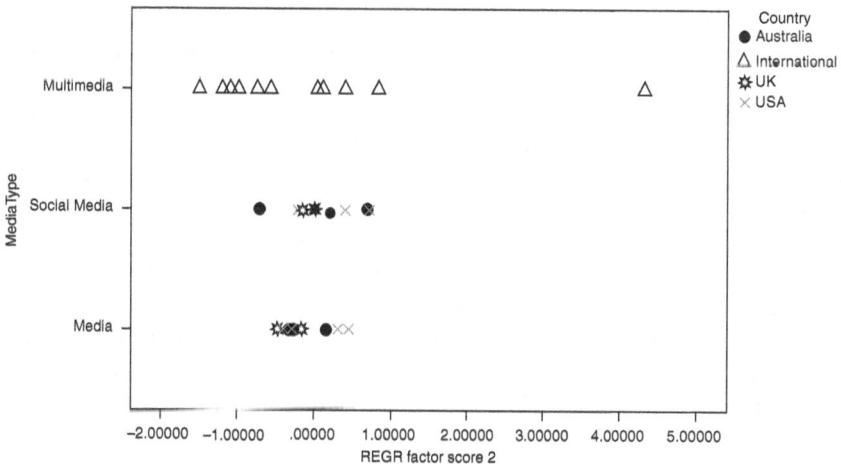

Figure 7.2 PCA Dimension 2: informal discussions of economic factors of China's air pollution.

Readers in the United Kingdom (represented by a star mark on the graph) who followed and responded to news reporting on the Chinese environment documentary seem to take a more formal communicative style, as the score of the UK dataset (0.12 for *The Daily Mail* and –0.14 for *Financial Times*) on this scale is much smaller. Finally, large variation has been found with the Australian social media data, as the informal speech scores attributed to the Australia new-based public posts vary considerably between 0.68 and –0.70. Similar patterns were observed with the media (news) category on the graph. US-based news reporting tend to use a more informal language style concerned with economic factors causing China's air pollution crisis: the *New York Times* (0.30); the *Wall Street Journal* (0.45); and the *Washington Post* (0.29). UK-based news, on the contrary, has negative scores on this scale: *The Daily Mail* (–0.18) and the *Financial Times* (-0.44) suggesting a more formal news language. Australian news is again polarized: ABC News (0.16); the *Sydney Morning Herald* (–0.25) and *The Guardian* (–0.32). Lastly, it was found that posts from viewers of the multimedia version of the documentary tend to use a more formal language (represented by a triangle mark), as most of multimedia data sets cluster on the negative side of the scale, with very few exceptions on the far right of the graph indicating a highly informal language.

Figure 7.3 shows the comparison between the news-based and multimedia-based social media in terms of the public reception of the Chinese environment documentary. The criterion of comparison is the third component of the five-dimensional PCA analytical model, which is emotion (see Table7. 2). A higher positive score on this dimension suggests a higher frequency of use of four linguistic and textual devices, that is, exclamation marks, quantifiers, second personal pronoun and adjectives. Typical social media messages, responses and posts which have high scores on the dimension of emotion are given in Examples 7.10–7.16. Figure 7.3 shows that most public posts on popular multimedia platforms (represented by dark pentagon marks on the graph) such as YouTube have positive emotional scores clustering on the right side of the scale. By contrast, public posts based on the news reporting of the Chinese environment documentary have negative scores, seemingly mirroring the less emotional tone and language of the news sources.

This is illustrated on the graph by the intermingling of news data (represented by circles) and news-based social media (dark triangles) on the left side of the scale. This corpus finding supports the hypothesis that viewers of the multimedia version of the original Chinese documentary with English subtitles are more emotionally involved than readers of news introducing and commentating on the Chinese environmental film. Another revealing finding consists in the use of the second person pronoun 'you' as a key linguistic marker of public posts

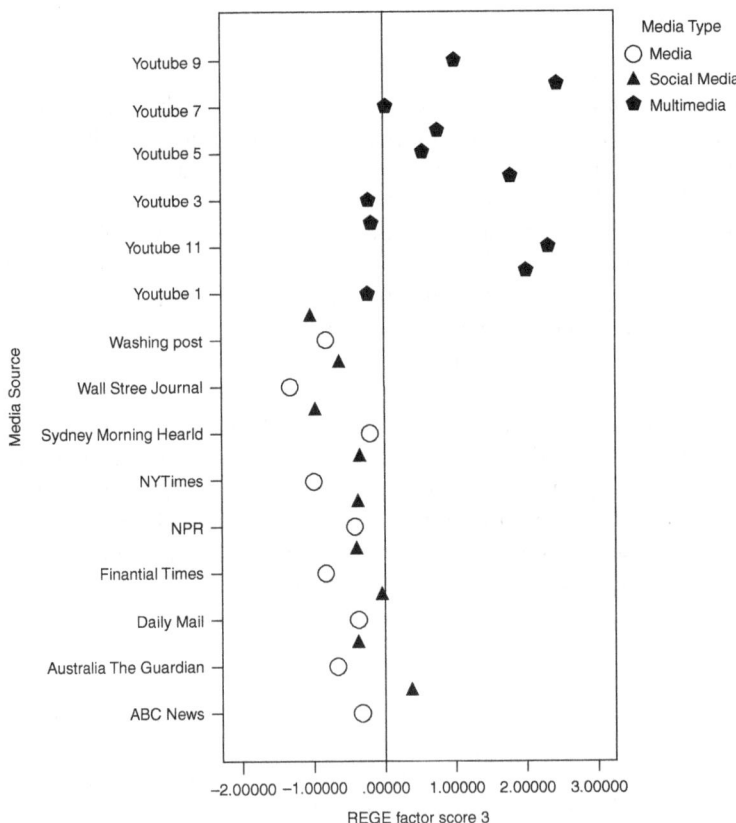

Figure 7.3 PCA Dimension 3: emotion.

which have high emotional scores. Examples 7.10–7.16 demonstrate that second person pronouns can have multiple references: sometimes, 'you' refers to fellow viewers or readers of the same news articles or YouTube multimedia contents; sometimes, 'you' is used to embody the Chinese state or administration. There are also cases when the second persona pronoun assumes the function of an indefinite reference, as viewers expressed strong personal emotion in reaction to the severe air pollution reported in the film.

Figure 7.4 illustrates how the gender and family relevance of environment health issues such as thee casual relation between air pollution and respiratory diseases was being explored in media and social media sources. First, in the multimedia-based social response category (represented by the top line of dark diamond marks), the majority of the data sets cluster to the left side of the scale indicating a low engagement with this topic among viewers of the multi- media versions of the Chinese documentary. This was largely replicated by the news-based social response category (represented by the middle line on the graph), as almost all of the public comments by readers of country-based news reporting on Chai's film have negative scores on this dimension. The finding points to a shared low engagement among English viewers and readers with the topic of the health impact of air pollution in China on children, women and family. However, the reporting, framing and critique of China's environmental crisis from the perspective of a mother and a female victim is the trademark and the winning reporting strategy of Chai's film, which holds the key to its wide social impact and success within China.

Despite the increase in the incidence of cancers and diseases related to poor air quality, before Chai's documentary, there had been very limited social debates and investigative news of the severe health impact of air pollution on young children and their families in China. Air pollution had been chiefly reported and exploited as an avenue of social tension between the government, industrial stakeholders and the public. Chai's reporting has drawn the public's attention for the first time to the health impact of air pollution on the life quality of vulnerable social groups such as children. Focus on improving the life quality of children and younger generations is perhaps the most important topic of many Chinese young families such as Chai's. Young parents born in the 1980s or after are mostly

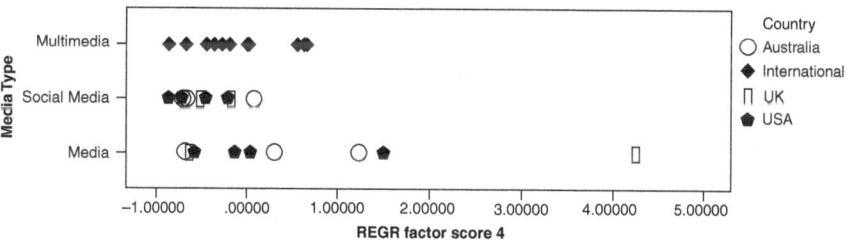

Figure 7.4 PCA Dimension 4: gender/family relevance of environmental health issues.

from Chinese families following the controversial one-child policy (1981–2016). The deprivation of quality life from early days of the children, – for example, in the case of Chai's daughter, the foetus was diagnosed with a benign lung tumour which the journalist and non-smoking working mother attributed to air pollution – can be devastating to young parents and their families. The missing link between the largely comprised life quality of China's younger generation and its environmental crisis was purposely explored and convincingly communicated to the public in Chai's investigative documentary. This highly important feature of Chai's film has been explored effectively and in depth in mainstream media. This is represented by the bottom line on the graph, with a number of newspapers clustering to the right side of the graph for their high positive scores of third scale of the gender and family relevance of air pollution. Newspapers that have heavily engaged with this topic are the ABC News (Australia) (1.24), *The Daily Mail* (UK) (4.27), NPR News (USA) (1.51) and the *Sydney Morning Herald* (0.32).

Figure 7.5 illustrates the use of words and expressions related to sensory experiences in news- and multi-media based social media messages. Positive scores on this scale indicate a higher frequency of use of expressions describing one's interaction with the subtitled audio-visual material of the Chinese environmental documentary such as seeing, hearing, feeling, expectation, and so on. Another important LIWC word category is leisure that includes a variety of media and social media sources such as Youku (Chinese version of YouTube); video games, TV series, tencent (Chinese multi-national conglomerate with subsidiaries specializing in various Internet-related services and products, entertainment, artificial intelligence and technology), Wechat (Chinese social media platform), Hollywood, and so on. Examples 7.20–7.25 show that some English-speaking viewers of the Chinese environmental documentary are keenly aware of the instrumental role of mass media and multimedia in promoting environmental protection and building shared understanding across cultures, peoples and societies.

Figure 7.5 shows there is important variation among viewers of the subtitled multimedia versions of Chai's film, with the highest score of 3.91 and the lowest

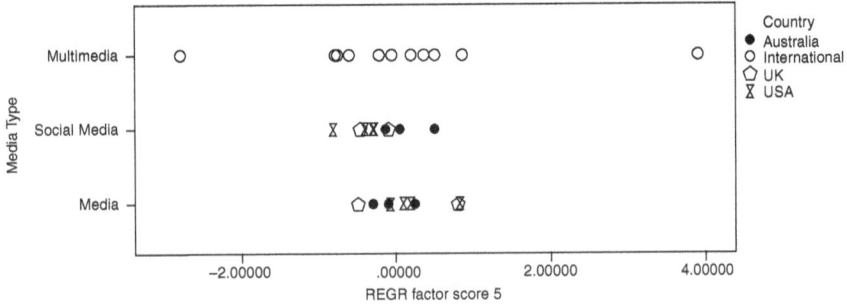

Figure 7.5 PCA Dimension 5: sensory experiences (seeing/leisure/hearing/feeling).

score of –2.79 (represented by circles of the top line). Among readers of news of the Chinese documentary, Australian readers seem to discuss more about the film based on their sensory experiences (represented by dark circles in the middle line clustering to the right side of the scale); and American news readers are least engaged with this topic. By contrast, US-based newspapers have large positive scores on this dimension (represented by funnel marks of the bottom line), pointing to a direct interest from American news reporters in the social impact created by the use of audio-visual materials and graphic resources in the media reporting of China's environmental pollution.

The following excerpt is taken from the *Wall Street Journal* sub-data set of the corpus constructed for the study:

> Chai Jing's film "Under the Dome," released on the Internet on Feb. 28, explained in graphic detail how all pervasive smog blackens lungs, poisons arteries and stunts lives. Alone on a stage, like Al Gore in "An Inconvenient Truth," she detailed how the coal and oil industries are complicit; how environmental officials are impotent. A giant screen behind her flashed data and video clips. The film quickly racked up 200 million hits. Then the censors moved in. First they instructed news media to play down the movie, then they banned it altogether. It stayed online in China for less than a week.
>
> (Pollution Film Too Popular for Beijing's Comfort:
> "Under the Dome" became a social phenomenon, threatening
> government's agenda, by Andrew Browne, March 17, 2015
> 12:38 AM ET, *The Wall Street Journal*).

7.7 Conclusion and future research

This chapter explored the social reception of Chai Jing's environmental documentary, widely considered as the Silent Spring in China. It adopted a broad definition of environmental translation to include both abridged written translation embedded in news reporting and the audio-visual versions of the original Chinese film with English subtitles. Using large amounts of news and social media data (approximately 200,000 words) from major newspapers in the United States, the United Kingdom and Australia and popular social media and multimedia platforms such as YouTube, this chapter analysed, configured and compared the different modes of the social reception of the Chinese documentary. Specifically, the corpus analysis identified five foci of discussions on mainstream newspapers and the social media which reflect the general picture of the public understanding and reception of the Chinese documentary on air pollution. These five foci or linguistic/text patterns of public discussions – extracted by using well-established corpus translation research methodologies such as principal component analysis – summarized as much as over eighty per cent of the media discussions of Under the Dome collected in the database. It was found that the most important pattern of public discussions around Chai's environmental film is a reflective thinking style characterized by a higher frequency of use of adverbs, verbs, auxiliary verbs, function words and expressions indicating cognitive processes and certainty; and a lower frequency of use of power-related expressions and words of six letters or more associated with more analytical and abstract thinking style.

This discussion pattern accounted for as much as 23% of the total media data collected in the corpus (Table 4). The second most popular pattern of discussion was identified as an informal communication style chiefly concerned with the economic factors of China's environmental crisis. The second communication pattern was widely seen in as much as one-fifth of the total discussions collected in the database. This pattern is associated with the use of linguistic devices such as expressions of non-fluencies, informal speech, money, anxiety and assent. The linguistic analysis found that this discussion style was preferred by major US-based news agencies such as the *Washington Post*, the *Wall Street Journal* and the *New York Times*. The third most important discussion pattern was summarized as the emotional tone of the viewers or readers of the documentary. This communication style was underscored by the use of exclamation marks, quantifiers, second personal pronoun and adjectives, and was found to be a defining feature of the comments made on popular multimedia platforms such as YouTube; but was not eminent in comments made by readers of news reporting on the Chinese environmental documentary. This may well be explained by the stronger emotional reaction elicited by the audio-visual versions of the original film with English subtitles, when compared with the more analytical and abstract introduction and reporting of the Chinese film in English news sources.

The fourth underlying communication pattern of the media materials collected in the corpus was described as a concern over the gender and family relevance of the health impact of air pollution in China. This pattern was found to be under-represented among social media users, but well-explored by professional news reporters in Australia and the United States. Lastly, the fifth key communication pattern was discussions around the sensory experiences of viewers of the film, and the instrumental role of multimedia in building a shared understanding of environmental protection among people from different language, cultural and social backgrounds. This pattern was found to be widely explored by viewers of the multimedia version of the Chinese film, and readers of news reporting on the film in Australia; and by professional news reporters in the United States. While the second, third and fourth patterns summarize and represent around one fifth of the discussion materials, respectively, the fifth pattern of public posts accounts for one tenth of the total media and social media data collected in the database.

The use of corpus methods identified five large patterns of the media communication and public comments which reflect the diversified social reception of the Chinese investigative environmental documentary among English-speaking audiences and viewers. The wide existence of these five large communication and commentating patterns was supported by numerous real-life examples taken from major English newspapers in the United Kingdom, the United States and Australia, as well as popular social media platforms like YouTube. The applicability and robustness of the five-dimensional analytical model developed in this study will be tested with media and social media data collected from other sources, countries and languages to advance current understanding of the role of digital media and translation as essential cross-cultural and cross-lingual communication tools in the promotion of a global environmental literature and culture.

References

Berry, D.S., Pennebaker, J.W., Mueller, J.S., & Hiller, W.S. (1997) Linguistic bases of social perception. *Personality and Social Psychology Bulletin*, 23(5), 526–37.

Dino, A., Reysen, S., & Branscomb, S.R. (2008) Online interactions between group members that differ in status, *Journal of Language and Social Psychology*, 28, 85–93.

Hook, G., Lester, L., Ji, M., et al. (2017) *Environmental Pollution and the Media: Political Discourses of Risk and Responsibility in Australia, China and Japan*, Abingdon: Routledge.

Newman, M.L., Pennebaker, J.W., Berry, D.S., & Richards, J.M. (2003) Lying words: Predicting deception from linguistic styles, *Personality and Social Psychology Bulletin*, 29, 665–75.

Xi, Jinpin (2017) Work Together to Build a Community of Shared Future for Mankind, speech at the United Nations Office in Geneva, Switzerland, Jan 2017 available at www.china.org.cn/chinese/2017-01/25/content_40175608.htm

8 The popularization of environmental issues in children's magazines

A cross-cultural corpus analysis[1]

Silvia Bruti and Elena Manca

8.1 Introduction

This chapter aims to investigate the strategies of popularization in texts about environmental issues addressed to children. Even though titles for young people have grown exponentially over the last few years, yet, to our knowledge, there are no comparative studies on the popularization of environmental issues for children. To the purpose, we have carried out several analyses, first comparing and contrasting texts written for children with texts written for adults on the same topic. Subsequently, since relevant differences have emerged for a variety of genres across languages, we have decided to compare popularizing texts for children in English and in Italian. In order to do so, we have identified suitable articles for a qualitative analysis and compiled corpora for a quantitative analysis. Apart from identifying specific linguistic devices and semiotic means that are used in popularizing texts for children, our aim is also to assess if these means correspond across languages, or, in other words, if the requirements of the genre are more influential than those of the language or vice versa.

The chapter is organized as follows. After briefly discussing non-fiction literature, we describe the strategies that need to be enacted when writing or adapting a text for young readers, given the different requirements they have in terms of encyclopedic knowledge, logical skills, attention, and so on.

Our qualitative analysis focuses on two comparable English texts, one for children and one for adults (respectively from *National Geographic Kids*, UK version, and *National Geographic* magazines), dealing with environmental issues, and a similar text aimed at children in Italian (from *Focus Junior* magazine). The quantitative analysis is based on five different corpora, one from *National Geographic* (NGAd), two from *National Geographic Kids* (both the American and British versions, e.g., NGKids_US and NGKids_UK), and two Italian corpora from *Focus* (FocusAd) and *Focus Junior* (FocusJun). The different variables, that is, readership and language, allow us to compare popularizing strategies aimed at different audiences and in different lingua-cultures.

8.2 Non-fiction literature for young readers

The relative lack of non-fiction[2] in literature for young readers was observed for the Anglo-speaking world by Coleman as far as 2007, but since then the number of informational texts written for young people has increased, with a consequent widening of the areas and topics included. As Colman herself rightfully underlines, apart from sustaining literacy in the same way as narrative texts do, non-fiction materials allow readers to acquire knowledge and skills that may orient their future jobs, and to become more informed and less naive citizens in many different spheres of life. Despite the growth in popularity of non-fiction, it is still rather surprisingly excluded from school programs and threatened by its big competitor, fiction, which is preferred by publishers because it is less costly (think for example of how high-definition photographs may impact on the final costs of non-fiction) and easier to handle, because it is 'fictive'. The two genres require in theory different approaches in reading, mainly an aesthetic stance for fiction, and an approach that pays attention to the logical concatenation of facts and to detail for non-fiction.

If the scarcity of information texts is still lamented in English-speaking countries, in Italy it has only recently gained notice, following in the footsteps of the Anglo-Saxon tradition, similarly to other ground-breaking genres.[3] LiBeR, an Italian quarterly publication on youth literature, publishes surveys detailing the genres that are issued yearly, comparing the data with those of previous years and from other European countries. In the last year surveyed, 2016 (www.liberweb. it/upload/cmp/Editori/LiBeR%20116%20-%20Rapporto.pdf; see Figure 8.1), fiction appears to have decreased compared to the previous three-year period (–1.2% in comparison with 2015), whose incidence (82.9%) remains however higher than in the entire previous historical series (the average from 1987 to 2016 was 77.4%, compared to 22.6 for non-fiction).

New non-fiction titles abound in three categories, which together cover more than half of the totality of printed books: nature, with 24%, followed by science/technology with 16% and thought/society 14% (see Figure 8.2).

When writing for children, some adjustments need to be made both in content and in language. Reading stimulates children's logical skills and provides them with information they can reuse. In order for a child to be able to understand and process the information that is contained in a text, the text itself needs to take into account the age and cognitive development of the addressee (see later in this chapter). Furthermore, the requirements of non-fiction books are different from those of fictional texts, which obviously relate to the plot, the characters, the setting. In non-fictional texts the first important requirement is linked to the accuracy of the information given (true but also devoid of stereotypes) and the way it is organized, for example from general to specific, from simple to complex. This movement is one that allows readers to identify the notions they need, to process them and gradually develop their thoughts. As Bianchi rightly argues (2018), the topic of how language should be molded to suit children's needs

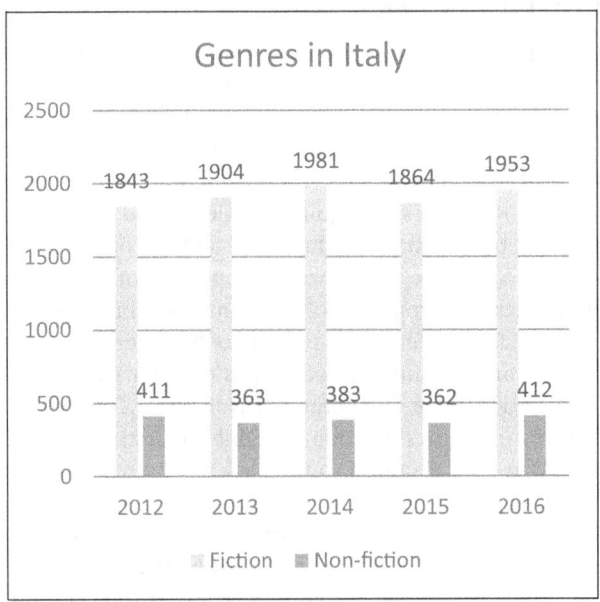

Figure 8.1 Distribution of fiction and non-fiction in Italy.

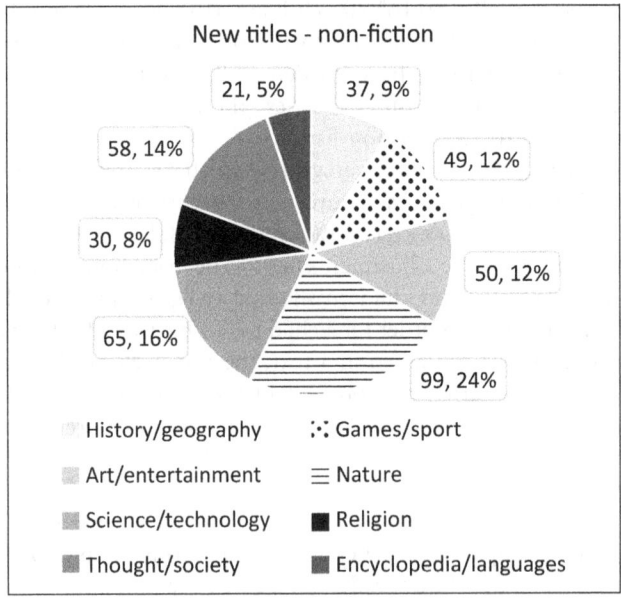

Figure 8.2 Genres within non-fiction in Italy.

(e.g., neither too simple or basic, understandable but with some unknown elements so as to stimulate the reader) is not dealt with in relation to texts for children but only in studies in language acquisition and reading comprehension.

When it comes to popularization, in recent years especially, the view has undergone revision and criticism (see Anesa & Fage Butler, 2015 for a useful review). Two points that have been raised in the discussion are extremely relevant when dealing with popularization for children: the very much debated problem that "popularization is a matter of degree" (Hilgartner, 1990, p. 528), that is, the border between expert and lay should be conceived as a gradient, does not apply in the case of texts for children, who are, in a certain sense, the purest example of lay readers. The other crucial issue, that popularization is a co-constructive process (Calsamiglia & van Dijk, 2004) in which changes occur in the roles taken by the actors involved, is confirmed by the engaging quality of popularizing texts for children, where addressees are immediately driven to use the acquired information to carry out a series of activities.

In line with previous studies on different genres of science popularization (see, *inter alia*, Motta Roth & Scotti Scherer, 2016), we hypothesize that popularizing texts for children are loosely interdiscursive, in that they combine features from the discourses of science, pedagogy and journalism. The qualitative analysis that follows in 8.4.1 aims at disclosing "the presence/absence of linguistic exponents of such interdiscursivity" (Motta Roth & Scotti Scherer, 2016, p. 173).

8.3 Strategies of adaptation to a child audience

Klingberg (2008, p. 12) argues that when texts not originally intended for children are adjusted so that they become suitable with regard to their real or assumed characteristics, they undergo a process of "adaptation" or "accommodation". Different types of adaptation can be applied: Klingberg (1970) suggests a division into matter-choosing, form-choosing, style-choosing, or medium-choosing adaptation. Furthermore, he identifies (1986) six macro-categories of adaptation: cultural context adaptation, modernization, purification, language adaptation, abridgement, and localization. These forms of adaptation occur not only during the translation process but also in case of interlanguage adaptation to a child audience.

In the translation of children's books, a role of utmost importance is played by the linguistic and the cultural adaptations (Klingberg, 2008), which, if not carefully handled, may contribute to the creation of a final product which is less interesting and less understandable than the original. Among the linguistic as well as the stylistic devices, there is readability connected to sentence length, which should be duly taken into account in that it may change from one language system to another. Added to this, the description of cultural concepts should also be carefully approached taking as a guideline the features of the target child audience and, consequently, their limited knowledge of the world. Shavit (1981, p. 171) argues that in the process of adaptation/translation, the translator of children's literature should always keep two basic principles in mind: 1. The text has to be adjusted in order to make it appropriate and useful to the child, in accordance with what society thinks is "good for the child"; 2. Plot, characterization and language need to be adjusted to the child's level of comprehension and his reading abilities. The first of these two principles is close to what Klingberg (2008) defines "didactizing" and "purification", two of the stages of the adaptation of a text he suggests to apply to children's literature. "Didactizing" is the introduction of an intention to instruct, which is an old tradition in this genre, whereas "purification" involves a type of adaptation which takes into account real or assumed sets of values of the addressees, both when the text is adapted for children and when it is translated for a different audience.

As mentioned above, adapting texts for children also implies the use of a number of linguistic strategies, which may contribute to make contents more accessible and interesting. Wrong choices in terms of lexis and syntax may lead to limited accessibility and confusion on the part of the children (Gamble & Yates, 2002; Reid & Donaldson, 1977; Barrs & Cork, 2001; Fox, 1993). Vocabulary choice should, for example, take into account age and level of knowledge of the children targeted in the text. For this reason, common words should always be preferred over technical or less frequent ones. The ability of children to perceive cohesive ties should also be considered. Chapman (1987) explains that this ability improves gradually with age and has a significant growth at an age ranging from 8 to 13. Of the four main cohesive groupings, cataphoric reference ties, ellipsis and conjunctive ties can be very confusing and demanding for children and, for this reason, need to be used carefully and sparsely (Gamble & Yates, 2002). Other syntactic constructions that may pose problems to children are hidden negatives, the passive voice, metaphorical statements and figurative language.

Bianchi (2018) has investigated the adaptation techniques used in two narrative versions of Shakespeare's *Romeo and Juliet* written in contemporary English for a young audience. Her results suggest the presence of a number of stylistic devices, which are very similar to those described in previous literature and briefly listed earlier. In the two texts considered for analysis, she notices a frequent use of very common words, repetition of specific phrases, a frequent explication of relations between circumstances and events in order to limit the use of cohesive ties, preference for overt negative structures, integration of dialogue into the narrative structure, use of concrete language and analogies to describe characters and their emotions.

As already anticipated earlier, the aim of our analysis is the identification of those strategies used by children magazines to describe and popularize environmental issues. According to Calsamiglia and Van Dijk (2004, p. 370), popularization is "a vast class of various types of communicative events or genres that involve the transformation of specialised knowledge into 'everyday' or 'lay' knowledge". This process of 'explanation' which allows language users to integrate their existing knowledge with a new lay-version of specialized knowledge may be operationalized through the following strategies (Calsamiglia & Van Dijk, 2004, p. 372ff): definition/description, denomination, reformulation or paraphrase, exemplification, generalization, and analogies such as comparisons and metaphors. Definition is a strategy used to explain unknown words while description aims to explain unknown concepts. Denomination is used when a specific concept, which is new to many people, needs to be introduced. It is a technical explanation, which helps readers understand the topic dealt with. The explanatory structures of reformulation and paraphrase are marked by relative clauses, appositions, parentheses, dashes, quotes and metalinguistic expressions, which aim to integrate old and new knowledge. Exemplification provides specific examples of general phenomena while generalization draws general conclusions from specific examples or cases. Metaphors and comparison are very frequently used strategies in that they are able to establish links between two domains of experience.

Diani and Sezzi (forthcoming) show that popularization strategies used in websites addressed to children disseminating information on the European Union are similar to those used in the recontextualization of expert discourse for a lay adult audience. The most frequent strategies identified are denominations, definitions, exemplifications, and similes and an overall tendency to simplification.

The aim of the analysis described in this chapter is, therefore, the identification of the popularizing and of the linguistic strategies which are used to adapt scientific information regarding environmental issues for a child audience. In order to do that, five corpora have been assembled and two types of analysis, a qualitative and a quantitative one, have been carried out.

8.4 Data analysis

To identify the main strategies at work in popularizing texts for children, in what follows we provide first a qualitative and then a quantitative analysis of popularizing texts designed for children (and compared to similar texts for adults), in English and Italian.

8.4.1 A qualitative analysis

The methodology used for the qualitative analysis is based on Discourse Analysis and on Multimodal Critical Discourse Analysis (Fairclough, 1992, 2003; Kress & Van Leeuwen, 1996, 2001). As announced in 2, we aim to identify the features of the various discourses interweaving in popularization articles for children. In particular, we expect to identify some devices that are shared with specialized written texts, such as accuracy, precision, orderly information, to sustain the informative function. Similarly, the texts are likely to accommodate devices with a pedagogical function, like reformulations and clarifications for clarity's sake, colloquial language, addresses to the audience, involvement strategies as well as traces of features of media journalism, such as the exploitation of diverse semiotics means, including the choice of typefaces and colors, and also cooperative tasks and the sharing of information on social media.[4]

For the qualitative analysis two comparable texts on the same topic were chosen: both deal with a crucial topic at present, that is, climate change, but one is expressly designed for children and one for adults. The two 'parallel' texts are available in the online versions of *National Geographic Kids*[5] and *National Geographic*. A brief comparison with an article from an Italian popularizing magazine for children follows.

The titles of the two articles are slightly different but the content is similar: the article for kids is entitled "What is climate change?"[6] and that for adults "Effects of Global Warming".[7] Usually, in texts destined for children, the topic announced in the title is presented as interesting and stimulating, but at the same time accessible and simple, so as not to discourage the audience from engaging in the reading. In this case the title is a question, announcing a quest for information in which the author involves his/her young readers, as is specified by the subtitle "we investigate ...". The pronoun form is evidently inclusive, signaling a path led by the author but followed by the audience. The title of the article for grown-ups is followed by a subtitle that zooms in on the points that are made in the body of the text, "The signs of global warming are everywhere and more complex than climbing temperatures". The subtitle points to the fact that a common belief will be somehow dismantled in the text and the so-much talked about topic of "global warming" is far from being a simple one. One of the typical ways of drawing attention to a topic by means of a title is make it sensational or multi-faceted and complex, so as to whet the readers' curiosity, a typical feature of the discourse of journalism.

In the *National Geographic Kids* article, the pronoun "we" used in the subtitle is maintained throughout the text (apart from the final paragraph, see below), together with the tendency to ask questions and provide answers. The article is organized in a very schematic way, with questions and answers, lists of points (either numbered or bullet-pointed), pictures to illustrate points, section titles and keywords in bold (e.g., **"What causes climate change?"** **"1. Burning fossil fuels"**, **"carbon dioxide"**) and clickable links in orange. Further emphasis is occasionally added by choosing to italicize some important words, not necessarily specialized ones, but words which the author wants to be focused upon (for example the pair of adjectives in the following sentence "The changing climate will actually make our weather more **extreme** and **unpredictable**").

The common pattern on which the text is built is the question/answer format (see Giannoni, 2008), which partakes two aims: it is a typical device of scientific discourse, but it also fulfils a pedagogic aim by involving the addressee. In providing answers, the author takes the opportunity of elucidating or defining concepts, explaining processes and often emphasizing unexpected effects of human behavior. The language used is mainly colloquial, as testified by the use of contracted forms ("don't just mean", "we'll get"), discourse markers ("so", "well", used in answers, as if they were turns in a conversation), vague terms and quantifiers ("stuff", "it means big things", "huge amount", the latter being specified later on, and also generic verbs like "make" "do" to replace more specific ones), formulae addressed to the readers ("believe it or not"), informal, everyday expressions ("it means big things", "cuppa", an assimilation for the noun phrase "cup of", "plus" used as an additive conjunction, "sure" used as an adverb in the sentence "it sure[8] adds up"), exclamatory remarks and exclamation marks (very often connected to the unexpected anecdotes told, e.g., "Unfortunately, rising temperatures don't just mean that we'll get nicer weather – if only!", or "They can learn to make fuel-efficient stoves which will not make them a little extra money, but also reduce the carbon footprint of the community – cool!"), and an occasional vulgar word, linked to scatological functions[9] ("fart").

When the author sets to explain the causes of climate change, he/she provides three reasons, each of which is treated separately and illustrated by a photograph. Apart from burning fossil fuels, the other two reasons provided for climate change are farming and deforestation, each of which has a devoted paragraph, including a picture followed by text. An image of factories with burning chimneys anticipates the content of the text, which aims at explaining the so-called 'greenhouse effect'. The strategy that is adopted here is that of employing a similitude, followed by the denomination of the scientific phenomenon: "The gases released into the **atmosphere** during this process act like an invisible 'blanket', trapping heat from the sun and warming the Earth. This is known as the '**Greenhouse Effect**'". As can be seen, a couple of words are highlighted in bold, in order to select the most important concepts in the text.

The second cause is farming, which is illustrated by a picture of cows in a field. In itself, the term "farming" is not particularly transparent and might be decoded as linked to agriculture and cultivations, but the picture more than the title reframes the content. The cause is explained in a very down-to-earth way, by saying that much of the methane gas that is in the atmosphere is due to digestive gases emitted by bovines, a piece of information presented as ludicrous, which is indicated by the use of suspension dots, the italicized word "fart", and an exclamation mark at the end. In the passage, there are several colloquial expressions, such as the address to the reader "believe it or not" and the use of "sure" as an adverb.

The third cause for climate change is deforestation. In this case, there is complete coherence between text and picture: there is in fact an image of a clear-cut area. Typically, specialized scientific terms are in bold, for example, **carbon dioxide**, together with keywords in the text, for example, **farmland**, **roads**, **oil mines** and **dams**, whereas the expression in orange, Amazon rainforest, is clickable, as it can be further explored at the reader's will. In this section there are two examples of defining strategies, the former applied to carbon dioxide, which is defined with an appositional explanation within dashes, and the latter related to the Amazon rainforest, for which a very vivid illustrative periphrasis has been employed, "the lungs of the earth".

The main paraphrastic devices used here are definitions and denominations, in line with what has been ascertained for scientific textbooks (see Bruti, 2004), which share with *National Geographic Kids* articles the aim of popularization but are geared to education rather than edutainment.

The article then proceeds to answering several other related questions regarding the ways in which climate change will affect the planet, wildlife, people and finally assessing how people can somehow counteract these effects. Towards the end, when talking about people, the article switches between different pronoun forms: although still employing the pronoun "we", the referent is specified as more restricted, for example, "we Brits love a good cuppa [...] but we probably take for granted just how much work goes into growing our tea", before alluding via the picture (a black woman seen from behind while gathering herbs that she collects in a basket that she carries on her shoulders) and the context to an implicit "they", referring to people living in developing countries. Finally, in the section that closes off the article, the focus is shifted onto the pronoun "I", in order to involve individual readers, make them feel responsible and push them to behave respectfully towards the environment. At the end of the article, there is a section for readers' responses, where children can leave their feedback by customizing their avatar using a selection of available images, have to write their name and add a comment. As stated earlier, in the body of the article there are several clickable words that suggest possible follow-up readings.

The online version of *National Geographic* offers a variety of articles devoted to the currently popular topic of global warming, one of which was issued on 30 January 2018. The article takes advantage of the multimodal potential of online journals by offering an almost three-minute long clip to better illustrate the title "What is Global Warming? The planet is heating up – and fast". The clip contains a clear and entertaining preview of what the article deals with, namely the causes and effects of global warming. The purpose of this introduction is to give a clear idea of the topic, anticipating some information, and raising the potential audience's curiosity. Although quite concise, the text of the clip is neatly organized into two parts, the causes and the effects of global warming; the former boil down to the greenhouse effect, which is the result of human activities throughout the centuries, whereas the latter are shown to have serious repercussions on four different dimensions, that is, oceans, weather, food and health. The contents explained by a female voice are elucidated by means of pictures, but also by graphics and data that show, for example, the increase of gases released in the atmosphere, and bulleted lists that allow the audience to visualize the elements while they are being described. The explanation of what scientists have discovered in the course of time about global warming is supported by the captioned picture of a polar bear, an iconic and widely recognized symbol of the disastrous consequences of the increase of temperatures and the ensuing threat to its habitat.

This article shares some features with the article for kids described earlier: the text, after an introduction and a paragraph devoted to the key notion of "greenhouse effect", is organized with questions and answers. Some of the questions are posed in section titles, "Aren't temperature changes natural?" "Why is this a concern?", and their answers are provided in the text that follows. Another similarity is that this article too underlines some concepts in deep yellow, the same hue of yellow that is used in both the *National Geographic* website and in the external edge of the print edition cover (which can thus be recognized as typical of the magazine by regular and occasional readers): the underlined words or expressions are clickable and once readers click on them, they are redirected to several available articles. In some cases they propose more detailed explanations of concepts (e.g., "melting", referred to glaciers, "rising", referred to sea level, etc.), whereas in others they offer accounts of topics that are touched here only in passing (for example the issue of "volcanic eruptions", which is more marginally linked with global warming). Some of the features of orality that are a distinctive mark of the article for kids are present here as well, although to a lesser extent. In particular, there are contracted verb forms ("it's becoming", "we've already"), the use of an inclusive "we", encompassing author, readers, but the humankind as well, and some utterance launchers that habitually occur in speech to take the turn or to pile one detail upon the other, namely "now" and "also". The choice of sparse and tempered forms of colloquialism is due to the fact that they need to appear appropriate for a scientific article destined for a grown-up audience. The degree of detail is also different, with precise figures being specified and supported by diagrams (in the video) and historical explanations being provided for various phenomena ("In 1895, the Swedish chemist Svante Arrhenius discovered that humans could enhance the greenhouse effect by making carbon dioxide..."). At the end of the article there are multiple suggestions for further reading, arranged in decreasing order of connection with the article at hand. While in the brief clip allusion is made to measures that can be taken to contrast the damaging outcome of the greenhouse effect to the environment, in the body of the article there is none, as the hyper-textual structure makes it possible and convenient to allocate space to this crucial issue in a separate article, that is however advertised in this page.

Some final words on the elucidating reformulation strategies utilized here. In popularizing texts, be they textbooks or magazine articles, the preoccupation is to inform, instruct, but also to make sure that concepts are clear and understood. In magazines, as has been seen, there is an additional preoccupation, that of making reading pleasant and entertaining, a major concern in texts for children, where special care is taken to use a complete array of devices to achieve the aim.

In the article under analysis there are three main paraphrasing strategies, two of which were also used in the article for kids, namely definitions and denominations. Denominations are useful with adults as well, in order to name concepts more precisely and technically: "by releasing heat-trapping gases as we power our modern lives. Called greenhouse gases...". Explanations instead provide the audience with definitions for specialized terms, as in "The 'greenhouse effect' is the warming that happens when certain gases in Earth's atmosphere trap heat". Another strategy that is used here is a kind of 'reductive' paraphrase, that is, exemplification. With this strategy, a specific example of a more general phenomenon is offered: one case in point is "Earth's remaining ice sheets (such as Greenland and Antarctica)", in which "Greenland" and "Antarctica" are examples of the superordinate term that precedes.

Although this qualitative analysis is not meant to draw generalizations, what seems to emerge is that the purposes of a genre, popularization in this case, entail using similar devices in texts that are thought for children and for adults. Multimodal strategies are exploited in both cases, using colors, fonts, hyperlinks, pictures or videos, but the edutainment nature is more evident in the text for kids, not only in attention to details and in devising involving activities, but especially in the choice of a colloquial language, one that is very close to the one children are exposed to in their daily lives.

The *National Geographic*, both for adult and for children, is published in translation in many countries, among which Italy. However, what we ideally wanted to investigate is the features of popularization for children in Italian, without the mediation of translation, which, as is well-known, may be responsible for several distorting effects. The most appropriate text for analysis seemed to be *Focus Junior*,[10] which is a monthly magazine for kids aged from 8 to 13, devoted mainly to science and technology, with an online edition. So we first carried out a search for the word "clima" ('climate'), and "cambiamento climatico" ('climate change'), but the results were too numerous and vague. So the second search was more specific, for example, "effetto serra" ('greenhous effect'), which refers to the content dealt with in the articles described above but employs a specialized term. The search gave a most appropriate result, an article authored by Giorgia Fanari, dating to August 2016. The presence of the name of the author is one difference between *Focus Junior* and *National Geographic Kids*, where no mention of the author is provided. The title of the article, "Che cos'è l'effetto serra" ('What is the greenhouse effect'[11]) suggests the same question/answer format used in *National Geographic Kids*, which is used several times in the body of the article. The similarities with *National Geographic Kids* are very many, namely, the explicit address to readers ("Siete mai entrati in una **serra**? Vi siete accorti del **caldo** che fa?" 'Have you ever walked into a greenhouse? Have you noticed how hot it is?'), the presence of questions, the organization of the argumentation in a bulleted list of points, each of which is supported by the presence of a picture, a combination of different fonts and colors, for example a bigger and blue font for the article subtitle, grey for photograph captions, bold for both paragraph headings and keywords in the text, and blue for clickable words which redirect to other articles. Another shared feature is that, despite the gravity of the situation and its consequences, both articles end on an optimistic note, with some measures that each of the young readers may take in order to contribute to limit the process and protect the environment ("**Nel vostro piccolo potete fare qualcosa per limitare le emissioni**", 'In your small way you can do something to reduce emissions'). Interestingly, the pronoun "you" in English and "voi" in Italian are employed to push readers to engage actively in environmental action. Likewise, at the end, there are some suggestions for further readings, divided into two groupings, a first one encompassing readings that children might like if they liked this article, and articles that are related to this.

As for the differences, some of them have to do with the more complex multimodal structure of the text. The title is followed by the subtitle and then by a picture, which explains how solar rays reach the earth, are partly absorbed and partly 'sent back', but blocked by greenhouse gases. The picture is quite complex, as it contains a graphic representation of this cycle, with several technical terms, but it is republished in the body of the article after the phenomenon has been explained. This picture is the first in a series within a slideshow. Another interesting difference is that pictures are captioned, mainly to provide their source, for example, Wikipedia or Pixabay. Furthermore, at the beginning and at the end of the article there are some evident traces of the social nature of the magazine: at the beginning, there are icons of the most known social media, so that readers can easily share this content with friends and acquaintances; at the end, there are several clickable hashtags for the article. All these elements obviously testify to the fact that the format has been purposely designed for young people, who are used to reading online and to sharing what they read with the various communities in which they belong.

From the point of view of the verbal material, the overall quality of the text is less colloquial in the Italian magazine. Although there is an attempt at making the text flow by means of questions addressed at the audience and a few conversational expressions (e.g., "insomma", 'all in all', and "un po' freddino, non vi pare?", 'a little coldish, don't you think?', with a use of the diminutive suffix *–ino*, employed in informal or child-centered situations, with intended pragmatic effects of non-importance, informality, etc.), the text is more formal than its English 'parallel' text. The presence of precise expressions (the nominalization "innalzamento", 'raising') and collocations ("raggi emessi", 'emitted rays'), the adoption of some subjunctive structures ("accordi che impegnino ogni Stato a contenere le emissioni", 'agreements that commit every state to contain emissions') make the tone of the Italian article more formal than that of the English one, where traces of orality abound in every sentence. As for typical popularization strategies, the tendency is to delve into the topic of greenhouse effect, which is explained by means of a typical definition, with the specialized term followed by a detailed explanation.

This qualitative analysis seems to suggest that in Italy care is taken in choosing engaging strategies to be used in popularizing genres, pivoting mainly around design and graphics, and availability and visibility on social media, but the verbal texture seems to be still dependent on the rhetorical tradition of the country.

8.4.2 A quantitative analysis

In order to carry out a quantitative analysis of the linguistic features of popularization in articles dealing with environmental issues, five corpora have been assembled and analyzed by adopting the methodology of Corpus Linguistics, and, in particular the keyword analysis. The articles included in the five corpora have been downloaded from the online versions of the US-based *National Geographic* magazine and *National Geographic Kids*, the British online version of *National Geographic Kids*, and from the Italian online edition of *Focus*, and *Focus Junior*.

The *National Geographic* magazine (www.nationalgeographic.com/magazine/) is one of the several media used by the National Geographic organization to captivate and entertain a global community focusing on issues regarding the exploration and the protection of the planet. Only those articles strictly dealing with the environment have been included in the corpus, that is to say, only those focusing on climate change and its immediate and future consequences, pollution, energy crises and new power sources.

The size of this corpus (henceforth NGAd) is of 15,588 word tokens and 3,569 word types. The Standard Type Token Ratio, which measures the relationship between the number of types and the number of tokens is 48.34. The higher the percentage of the STTR, the more varied the vocabulary. This figure may be better interpreted if compared with the STTR of the British National Corpus (BNC) spoken and written sections. The spoken section of the BNC has a low STTR, 32.48, which is typical of those texts, which do not have a varied vocabulary. The written component of the BNC has 57.81 as STTR, which accounts for a higher linguistic complexity. The NGAd thus shows to be closer to a written genre, although it does not have a very high linguistic complexity.

The US online version of the *National Geographic Kids* (https://kids. nationalgeographic.com/) aims to "teach kids about the world and how it works, empowering them to succeed and to make it a better place".[12] It provides children with videos, games, instructions on how to carry out experiments at home, a section dedicated to 'Homework Help' where animal and country profiles as well as encyclopedia entries are provided, an interactive map, and several links to useful information on the planet and its treasures. The articles downloaded from the online version of the magazine feature tips and facts on planet Earth, climate change and pollution. This corpus (hence NGKids_US) has 5,707 words and 1,466 word types. Its STTR is 45.16, a figure that may account for a less varied and more repetitive vocabulary with respect to the version for adult readers.

The British version of the National Geographic Kids (www.natgeokids.com/ uk/) has a main section called 'Discover section' where children can discover "some of the coolest facts for kids from around our planet". It is further divided into four groups, 'animals', 'geography', 'science', and 'history'. Most articles contain interesting and curious facts about people, places and animals as suggested by the titles '15 fun facts about the Queen', 'Facts about the moon', '10 facts about pandas'. The articles downloaded from this website and included in our corpus focus on facts about planet Earth's features and places (such as rainforests, oceans, volcanoes, hurricanes, the Arctic), climate change issues and conservation tips, energy crisis, pollution. The total number of word tokens in this corpus (henceforth NGKids_UK) is 6,699 and it has 1,642 word types. Its STTR is 48.03, a figure that may lead us to hypothesize that its vocabulary is highly varied.

The two Italian magazines selected for analysis are the online versions of *Focus* and *Focus Junior*. *Focus* (www.focus.it) is a monthly magazine which defines itself as "the point of reference of scientific popularization and entertainment with a clear and appalling style which emotions and excites". The online version is divided into different sections, such as science, environment, technology, culture, behavior, and links to a picture gallery and to other media, such as a TV channel, which belongs to the *Focus* group. For the present analysis, only those articles related to science and environment (such as, climate change, pollution, renewable power sources, sustainable food) have been downloaded in order to make this corpus comparable to the three English corpora (NGAd, NGKid_US and NGKids_UK) and to the Italian corpus of children magazine articles (FocusJun). The size of this corpus (henceforth called FocusAd) is of 11,666 words and 3,099 word types. The STTR is 44.04 but, unfortunately, to our knowledge, statistical measures of an Italian general reference corpus do not seem to exist or to be available.

Focus Junior (www.focusjunior.it) is a monthly magazine belonging to the Focus group. It is specialized in scientific popularization for children aged between 8 and 13. It aims to "satisfy and stimulate children's innate curiosity and to entertain them by offering a new and intelligent way of discovering the world". The articles downloaded from this website focus on pollution and recycling, natural phenomena (such as earthquakes, storms, aurora borealis, volcanoes, and so on), and climate change. This corpus (FocusJun) has 8,666 running words, 2,639 word types, and a STTR of 43.8, which is quite similar to the figure identified for the FocusAd STTR. This may lead us to hypothesize that the two corpora have a similar linguistic variety although they target two different age groups.

The first step in the comparison of the three English corpora and the two Italian corpora is the creation of keyword lists by using AntConc,[13] a software package for concordancing and text analysis. The tool Keywords shows how distinctive a word is: when 'keyness' (i.e., a percentage indicating unusual occurrence) is very high, the word can be called a keyword, that is to say a word, which occurs more often than would be expected by chance in comparison with a reference corpus (Scott 1999, p. 236). The importance of a word, thus, strongly depends on its frequency. Keywords can be of various types (Scott 1999, p. 71): proper nouns, content words which account for the 'aboutness' of the text, and function words, which are indicators of style. Function words are usually high-frequency words and a manual analysis would hardly identify them as keywords. For this reason, their presence in the keyword list is highly unusual and worth investigating. By comparing the wordlist of the NGAd with those of the NGKids_US and NGKids_US we may identify those words which are peculiar of the children's magazines and, for this reason, may be considered as preferred popularizing strategies for a child audience.

8.4.2.1 *NGAd, NGKids_US and NGKids_UK*

Let us start our analysis by comparing the NGAd with the NGKids_US. The word with the highest keyness is "your" (keyness value: 161.340), but a number of interesting words show up until the keyness value 13.175. Furthermore, the occurrences of these key items has also been calculated in terms of percentages in order to make the comparison across the two corpora easier.

The most interesting key category is that of personal pronouns: the pronouns "you" and "your", which are top of the keyword list, occur in the NGKids_US with a percentage, respectively, of 1.17% and 1.45%. Indeed, in the NGAd, these pronouns seem to be less used as percentages suggest: "you" occurs with a percentage of 0.23% and "your" with a percentage of 0.06%. Conversely, the pronoun 'we' shows to be more used in the NGAd than in the NGKids_US: it occurs with a percentage of 0.53% in the NGAd and of 0.19% in the NGKids_ US. Interestingly, the personal pronoun "our" is slightly more frequent in the NGKids_US (0.22%) than in the NGAd (0.14%).

The preference for the second person pronoun in the NGKids_US corpus may be considered a popularizing strategy that aims to involve the young readers by addressing them directly. Indeed, although the pronoun "we" is used in an inclusive way, children may have difficulty in recognizing themselves in the people this pronoun alludes to. Conversely, a direct address may sound as a call to action and, consequently, maybe more stimulating and appealing.

The didactic purpose, which is a typical popularizing strategy, of the article may be seen in the keywords "instead", "tips", "ask", and "can", which mainly refer to what children should and can do for their planet. Their percentages of occurrence in the two corpora confirm our hypothesis (see Table 8.1).

Some of the keywords in the list are examples of informal vocabulary, such as "trash", "cool", and "lots". Their percentages of occurrence are namely: 0.38%, 0.07%, and 0.07%. While "trash" is also used in the NGAd (0.04%), the other two items show no occurrences. The preference for common/informal words over more technical words is another popularizing strategy, which aims to make the text more suitable to children's level of knowledge and, for this reason, more interesting.

An interesting keyword is the conjunction "or", which occurs with a percentage of 0.91 in the NGKids_US and of 0.32% in the NGAd. It is frequently used to give an alternative explanation, description or suggestion about an issue as in "Animals might have been harmed or disturbed to make them" or in "Try to avoid buying products that use it, or look for a label that confirms the ingredient". For this reason, it may be considered a further popularizing strategy used in the children's magazine.

In order to have a further confirmation of our hypotheses regarding the popularization strategies used for a child audience, we will compare the NGAd against the NGKids_UK.

The items selected in the keyword list have a keyness which goes from 71.126 to 10.655. As in the previous analysis, at the top of the list we find the personal pronouns "you" and "your". "You" (1.18%) is much more preferred to "we" (0.53%) in the NGKids_UK, while "our" (0.40%) is slightly more preferred

Table 8.1 Percentages of words used with didactic aims in the NGAd and the NGKids_US

	NGAd	*NGKids_US*
INSTEAD	0.02	0.35
TIPS	0	0.14
ASK	0	0.21
CAN	0.39	1.06

to "your" (0.33%), exactly as it happened in the NGKids_US. This tendency confirms that the use of the second-person pronoun to address directly the young readers is a typical popularizing strategy in children's magazines. The preference for "our", instead, could be interpreted as a strategy of involvement, which aims to make children feel as part of a big community.

The modal "can", occurring with a percentage of 0.92%, has a similar usage as that identified in the NGKids_US, that is to say telling and describing what is possible and what children can do. However, in the keyword list, we do not find verbs whose purpose can be considered as didactic. Conversely, the presence of "join" (0.10%), "check" (0.12%), and "imagine" (0.12%) suggests the strategy of involvement, of a call to action which aims to stimulate and engage the young readers more. These verbs have no occurrence in the NGAd. The search for involvement is also visible in the use of informal language, as suggested by the items "super" (0.10%), "cool" (0.07%), "fascinating" (0.07%), and "gang" (0.08%). These items as well have no occurrence in the NGAd.

To sum up, from these key items it can be assumed that the language adaptation strategies of both the British and the American versions of *National Geographic Kids* are based on the following devices:

- preference for the personal pronoun "you" to create higher interaction and involvement with the children, and of the pronoun "our" to have the young readers perceive they are part of a big community;
- preference for an informal language, which may make contents clearer, less confusing, and more entertaining;
- use of imperatives with a didactic purpose or to stimulate readers and involve them more;
- use of items, such as "or" (but only in the NGKids_US), which allow the writer to explain concepts better and in a clearer way by adding further descriptions and definitions.

In the following section, the analysis proceeds with the two Italian magazines. Our aim is to check if popularizing strategies are universal or depend on the linguistic and cultural systems.

8.4.2.2 FocusAd and FocusJun

The keyword list obtained comparing FocusAd and FocusJun is very interesting, in that it includes key items, which are different from those identified in the previous analyses.

The strong tendency towards the use of the second-person pronoun seems to be absent in the list, although the fact that Italian verbs are inflected may contribute to make this difference not visible. Further analyses should be conducted to state that. Conversely, what strikes most is the presence of adverbs and conjunctions, which have either the role of catching the young readers' attention, or to explain and exemplify. Words belonging to the first group are: *ecco* ('here it is/here you are'), performing the function of drawing the readers' attention, and occurring with a percentage of 0.20% in the FocusJun but absent in the FocusAd; and *proprio* ('exactly/precisely'), which is used to insist on the concept of the word it determines. Key items belonging to the second group and, for this reason, used to exemplify and make connections clearer are: *come* ('like/such as'), used to introduce a simile, an explanation or a way something works or acts; *perché* ('why/because') used to introduce a clarification or to ask a question; *infatti* ('in fact/indeed'), which is usually followed by something which serves as a confirmation or as an example of what is being talked about; *ossia* ('that is'), a conjunction which introduces a clarification or a description; *esempio* ('example'), used in almost all cases in the phrase *per esempio* ('for example') introducing an exemplification; and *invece* ('instead'), an adverb used to compare and contrast two concepts thus making their features clearer, as in ... *la parola solstizio viene dal latino Solis statio: fermata, arresto del Sole. Il primo giorno di primavera è invece detto equinozio* ('the word solstice derives from the Latin *Solis statio*, meaning stop of the Sun. The first day of spring, instead, is said equinox').

The modal *potere* and its conjugated forms are also indicated as key items, but they are used more frequently to refer to activities, events, and consequences that can happen and occur, differently from the English corpora, where they were preferably used to tell children what they can or should do.

To sum up, the popularizing strategies that seem to be more frequently adopted in the Italian children's magazine mainly focus on involvement and explanation through the use of adverbs and conjunctions, which emphasize and explain concepts. Furthermore, young readers' involvement seem not to be achieved through informal language or vague language as in the two English corpora.

8.5 Conclusion

The results of the quantitative analysis confirm what already identified in the qualitative one, that is, that the popularizing articles in science magazine for children reveal "the functioning of interdiscursivity among the discourses of science, journalism and pedagogy" (Motta Roth & Scotti Scherer, 2016, p. 190). In the *National Geographic Kids* magazines, both the British and the American versions, popularizing strategies for a child audience aim to involve, entertain, and educate the reader. In order to do this, language and visuals support each other in the explanation of concepts, facts and events. In fact, along with the use of pictures, graphic devices help readers focus on definitions, names, and terms. The English magazines tend to involve their readers through a question and answer format, which is more reader-friendly and contributes to catch the young readers' attention, who would probably be confused or less interested by a long and single paragraph. Furthermore, concepts are expressed in a colloquial language, using vague terms and informal expressions, a feature which shortens the distance between the writer and the reader. For this reason, the type of interaction created by means of the text seem to be more peer-to-peer rather than expert to non-expert. A further confirmation of that is the prevailing use of the personal subject pronoun "you" and of the personal possessive adjective "our", which well balance interaction and involvement, as if readers were called to action to get to know and save our planet.

The Italian children's magazine also recurs to the question and answer format to make contents more stimulating and interesting and to catch the young readers' attention. However, the language used does not seem to undergo a process of adaptation with respect to the adult version. Indeed, adaptation strategies are mainly content-based and focus on the process of explanation, definition, and description as the presence of key adverbs and conjunctions confirm. The rhetorical tradition of the country, with a marked preference for a written style that is separated from the oral, is, therefore, visible in these linguistic choices. Young readers are left the opportunity of using a more informal language or being more active when writing comments on articles or when sharing them on the socials which are mostly used by Italian children.

Although further analyses would be needed to generalize on the features of popularization in the two languages/cultures, it may be assumed that the requirements of adaptation to a child audience are culture-dependent. The popularizing strategies used in the three corpora from children's magazines contain the same popularizing strategies described in the literature but the English and the Italian corpora tend to use them with different frequencies. This accounts for two diverging approaches to a young audience and to the notion of edutainment.

Notes

1 The research leading to this contribution was conducted by both authors. Silvia Bruti wrote sections 8.1, 8.2 and 8.4.1; Elena Manca wrote sections 8.3, 8.4.2 and 8.5.

2 The two alternative labels that are used to refer to popularizing texts for children are children's non-fiction texts or informational texts. Mallett (1999), for example, adopts the term non-fiction and further distinguishes between narrative non-fiction and non-narrative non-fiction, whereas Vardell (2014) claims to prefer the definition of informational texts. The dichotomy fiction/non-fiction is a false one and might wrongly induce children into believing that fiction books are not read for information, when in fact fiction books are an important source of information that has been carefully researched and inserted within a narrative form; and that non-fiction books do not entertain. What is by now largely shared is that these texts attempt to instruct and amuse at the same time. This recent trend is reflected in their 'hybridity', e.g., texts that are mostly factual but contain invented characters and material which have therefore been labelled either as "blended" or "faction" (see Colman, 2007, pp. 160–1).

3 See for example the recent spread of the genre of tourist guides for children from English into Italian. Cappelli and Masi (forthcoming) claim that "It is, however, also possible, that, in order to fill a void in the market, the Italian publishing industry has taken inspiration from well-established editorial products available in the English-speaking countries or commissioned translations of English materials".

4 The spread of social broadcast technologies has favored a shift in news (or information) consumption and production: consumers are no longer passive "receivers" but

actively participate in creating and exchanging contents. For a review of the various effective uses of social media, see Henderson, Snyder, & Beale (2013).

5 This article belongs to the UK-based version of *National Geographic Kids*.

6 www.natgeokids.com/au/discover/geography/general-geography/what-is-climate-change/

7 www.nationalgeographic.com/environment/global-warming/global-warming-effects/

8 The term "sure" is certainly a marker of orality when used in writing, giving the text the flavor of informal speech, Examples of these uses are "She sure dazzled the audience with her acceptance speech", or "It was sure hot enough in the auditorium", both retrieved at www.urbandictionary.com.

9 Both fictional and non-fictional authors are aware of children's liking for macabre, profane and filthsome details, including the so-called 'potty humour', all of which were for instance skillfully concocted in Roald Dahls' tales (consider, his masterpiece of subversion of popular fairy tales, *Revolting Rhymes*, 1982; see Anderson 2016).

10 *Focus* is a monthly magazine on science, sociology and current affairs that is published in many countries. The Italian edition, published by Mondadori, was born in 1992. The magazine explores topics from the world of science and current affairs, but occasionally it also offers some space to opinion polls on current issues. Like the *National Geographic*, *Focus* has a version for children, which is called Focus Junior (see 8.4.1 for further details). Like all print media, *Focus* and *Focus Junior* have a web edition and a TV channel.

11 www.focusjunior.it/scienza/ambiente/che-cos-e-l-effetto-serra/

12 https://kids.nationalgeographic.com/about-us/

13 AntConc is freely available at www.laurenceanthony.net/software/antconc/

References

Anderson, H. (2016). The dark side of Roald Dahl, 13 September 2016, BBC Culture. Retrieved from www.bbc.com/culture/story/20160912-the-dark-side-of-roald-dahl.

Anesa, P., & Fage-Butler, A. (2015). Popularizing biomedical information on an online health forum, *Ibérica*, 128(29): 105–28.

Barrs, M., & Cork, V. (2001). *The reader in the writer*. London: Centre for Language in Primary Education.

Bianchi, F. (2018). Rewriting *Romeo and Juliet* for a young audience. A case study of adaptation techniques. Lingue e Linguaggi (special issue), 27, 43–65.

Bruti, S. (2004). La voce dell'autore. Alcuni tipi di parafrasi in un corpus di inglese scientifico. In S. Bruti (ed.), *La parafrasi tra messa fuoco del codice e negoziazione discorsiva* (pp. 89–109). Monographic number of *Rassegna italiana di linguistica applicata, XXXIX* (1). Roma: Bulzoni.

Calsamiglia, H., & van Dijk, T. A. (2004). Popularization discourse and knowledge about the genome. *Discourse and Society*, 15(4), 369–389.

Cappelli, G., & Masi, S. (forthcoming). Knowledge dissemination through tourist guidebooks: Mind the gap between adults' and children's. In M. Bondi, S. Cacchiani, S. Cavalieri (eds), Communicating specialized knowledge: Old genres and new media. Newcastle upon Tyne: Cambridge Scholars Publishing.

Chapman, J. (1987). *Reading: From 5–11 years*. Milton Keynes: Open University Press.

Colman, P. (2007). A new way to look at literature: A visual model for analyzing fiction and nonfiction texts. *Language Arts*, 83(3), 257–268.

Dahl, R. (1982). *Revolting rhymes*. London: Penguin.

Diani, G., & Sezzi, A. (forthcoming) The EU for children: A cross-linguistic study of web-mediated knowledge dissemination. Paper presented at the Clavier International

Conference "Representing and Redefining Specialised Knowledge", Bari 30 November–2 December, University of Bari.

Fairclough, N. (1992). *Discourse and social change.* Cambridge: Polity Press.

Fairclough, N. (2003) *Analysing discourse: Textual analysis for social research.* London: Routledge.

Fox, C. (1993). *At the very edge of the forest: The influence of literature on storytelling by children.* London: Cassell.

Gamble, N. & Yates, S. (2002). *Exploring children's literature. Teaching the language and reading of fiction.* London: Paul Chapman.

Giannoni, D. S. (2008). Popularizing features in English journal editorials. *English for Specific Purposes,* 27(2), 212–232.

Henderson, M., Snyder, I., & Beale, D. (2013) Social media for collaborative learning: A review of school literature. *Australian Educational Computing,* 28(2), 1–15. Retrieved from http://journal.acce.edu.au/index.php/AEC/article/view/18.

Hilgartner, S. (1990). The dominant view of popularization: Conceptual problems, political uses. *Social Studies of Science,* 20(3), 519–539.

Klingberg, G. (1970). *The fantastic tale for children: A genre study from the viewpoints of literary and educational research.* Gothenburg: Gothenburg School of Education.

Klingberg, G. (1986). *Childrens' fiction in the hand of translators.* Malmo: CWK Gleerup.

Klingberg, G. (2008). Facets of children's literature research: Collected and reused writings. The Swedish institute for children's books, no 99. Retrieved from www.sbi.kb.se/upload/Public/Forskning/Klingberg_Facets.pdf.

Kress, G., & Van Leeuwen, T. (1996). *Reading images: The grammar of visual design.* London: Routledge.

Kress, G., & Van Leeuwen, T. (2001). *Multimodal discourse: The modes and media of contemporary communication.* London: Arnold.

Mallet, M. (1999). *Young researchers: Informational reading and writing in the early and primary years.* New York: Routledge.

Motta Roth, D., & Scotti Scherer, A. (2016). Science popularization: Interdiscursivity among science, pedagogy, and journalism. *Bakhtiniana,* 11(2), 171–194.

Reid, J., & Donaldson, H. (1977). *Reading: Problems and practices.* London: Ward Lock.

Scott, M. (1999). *WordSmith Tools Help Manual. Version 3.0.* Mike Scott and Oxford University Press.

Scott, M. (1997). PC analysis of key words – And key key words. System, 25(2), 233–245.

Shavit, Z. (1981). Translation of children's literature as a function of its position in the literary polysystem. *Poetics Today,* 2(4), 171–179.

Vardell, S. M. (2014). *Children's literature in action: A librarian's guide* (second edition). Englewood, CO: Libraries Unlimited.

9 Anglicisms in Italian environmentally friendly marketing

English as the global language of capitalism or sustainability?

Maria Cristina Caimotto

9.1 Introduction

The aim of the present chapter is to investigate the role of English as a Lingua Franca in international communication concerning environmental issues. The analysis concentrates both on the use of Anglicisms in other languages (focusing on Anglicisms in Italian) and the use of English as the common language for international environment-related communication. In order to do so, it is necessary to start from a reflection on the role played by English in the world of global marketing and business communication.

It is a fact that in history no other language has ever been as widespread as English is nowadays: this can either be welcomed as a new possibility for humanity or as proof of dangerous linguistic imperialism. These antithetical views have been the object of study by many scholars: this paper refers to the works of Crystal (2003) – as representative of a mainly positive attitude towards English as a global language – and those of Phillipson (1992), who argues against the mechanisms of power hidden in the concept of a *Lingua Franca*.

Crystal identifies the main historical reasons that made and conserved English as the global language: in the 19th century the supremacy of Britain as "the world's leading industrial and trading country" and its political imperialism, and later the economic hegemony of the new American superpower in the 20th century (Crystal, 2003, p. 10). Hence, industrial growth and the stock market are the elements that have helped English establish its supremacy. It has been argued that English is the language of capitalism (Block, Gray and Holborow, 2012; Fairclough, 2006) and, as a consequence, one could analyse it as the language of power and the cultural symbol that represents the approaching environmental breakdown. But this view would prove simplistic, and this paper aims to investigate other and antithetical cultural symbolisms associated to the English language.

The same clash of views appears to be present in the use of Anglicisms in Italian; hence contrasting hypotheses are presented here in order to investigate their respective validity and to understand how they coexist and what can be done in terms of discourse strategies in order to improve environment-friendly communication.

The first hypothesis is that in some cases the presence of Anglicisms can be considered as an alert to greenwashing strategies, or at least some non-totally transparent green marketing strategy (see Caimotto and Molino, 2010). In other cases, the use of English is a strategy that helps companies promoting their business activities that really are environmentally friendly (Caimotto, 2013). We must remember the older symbolisms of English as the language of political liberty; hence the second hypothesis takes into account the positive communication potential of Anglicisms (Crystal, 2003). Meanwhile it is also important to bear in mind that in some cases the use of English is dictated by a general identification of English as the language of business and marketing, without prior planning or ideological goals (Stubbs, 2001). This kind of usage, though, is also likely to engender implications and specific framing (Lakoff, 2010): it is the aim of this work to identify and analyse such implications together with their effects and to elaborate strategies that might lead to a more conscious and focused use of Anglicisms and of English as a global language, as far as green communication is concerned.

In order to achieve this result, several communicative situations are taken into account and analysed, starting from the observation of the use of Anglicisms in different Italian contexts. Drawing on previous research (Caimotto and Molino, 2010; Caimotto, 2013) the aim here is to create a taxonomy of environment-related communicative practices in order to identify their characteristics and establish what works and what does not in terms of both effectiveness and transparency.

The first case study focuses on various companies that took part in a national fair held yearly in Milan "Fa' la cosa giusta" ("Do the right thing") which describes itself as the fair "of critical consumption and sustainable lifestyles," at its tenth edition in 2013. Then another case study is presented: that of Termoindustriale, a company based in Northern Italy operating in the business of cogenerators, also known as CHP (combined heat and power). All these situations involve the contradictions discussed above, as they require marketing strategies, raise environment-related issues and imply various degrees of economic interest.

The methodology employed for this investigation is qualitative. Corpus-aided approaches can be a valuable tool when carrying out critical discourse analysis (see Baker et al., 2008) and some of the works investigating the discursive role of Anglicisms in Italian have employed corpus-aided approaches (e.g. Fusari, 2012; Caimotto and Molino, 2010; Boggio, 2017) but there are cases in which a qualitative approach can complement the findings from corpus based studies. As illustrated in the following sections, the analyses presented here for the first case study were constructed through qualitative observation inside a fair. This would correspond to the ordinary experience of a visitor to the fair and, as such, the investigation complements a corpus-based one, which analyses an amount of text a single person is unlikely to read. Moreover this work focuses on the role of Anglicisms in Italian and, as Furiassi (2008) explains, there are drawbacks when employing automatic procedures to retrieve them.

Boggio (2017) and Furiassi (2017) also demonstrate that in order to observe the nuances of the influence that Anglicism can have upon Italian discourse, the best solution is to approach the texts from a qualitative perspective which can then be completed through corpus-aided quantitative investigations. In the second case study analysed in this chapter, a specific company website is observed closely with the aim of showing the role that a business-oriented discourse – characterized by the prominent presence of Anglicisms – can play in a situation in which the importance of business over environmental issues is at stake.

9.2 The myths of English

The implications of the use of English as the language of communication are engendered by the narratives associated to the language and to the cultures forming the Inner Circle (Kachru, 1988). To these, we must add the culture of globalization and the concept of English as a *Lingua Franca*. Depending on which narrative we are referring to, the English language can represent either a symbol of environmental and ecological destruction or a symbol of rural idyll and democratic ideals. In order to understand how such contrasting narratives can coexist in a whole, Lakoff's work (2010) on framing the environment can prove useful.

According to Lakoff, the ways in which we conceive the environment suffer from severe hypocognition (p. 76), that is, we lack the ideas we need in order to communicate environment-related issues effectively. In Lakoff's words:

> *The economic and ecological meltdowns have the same cause*, namely, the unregulated free market with the idea that greed is good and that the natural world is a resource for short-term private enrichment.
>
> (p. 77)

A company that foregrounds its economic interests to the detriment of its environmental impact is to be considered on the wrong side in terms of green credentials. But a company cannot be entirely oblivious of the economic sustainability of any of its projects. Hence, what a critical consumer needs to do is to understand whether the company's economic interest will be placed before the environmental issues or not. Unfortunately, this requires good technical knowledge that most lay people do not have. Thus, people who care about the environment sometimes make the mistake of judging a company negatively simply because it will earn good money from its environmentally friendly business.

Lakoff writes about the new frame of The Regulated Commons as a possible solution to improve the issue of hypocognition. The Regulated Commons consist in "the idea of common, non-transferable ownership of aspects of the natural world, such as the atmosphere, the airwaves, the waterways, the oceans, and so on" (Lakoff, 2010, p. 78). As the idea of the Commons was an important aspect of the Magna Carta (Linebaugh, 2008), we see a further connection to English-speaking culture.

Still, we are also well aware of the cultural role played by the United States as the country of capitalism, which – as explained by Lakoff – is directly connected to environmental issues. English is considered the language of international business communication and several scholars point out the negativity associated to this language. Among the most well-known we find Phillipson (2008), who argues that English is a *lingua frankensteinia* rather than a *lingua franca*, and certainly not as neutral and value-free as the adjective *franca* appears to suggest. Likewise, Dieter (2004, p. 140) argues that the language of marketing is what he labels as BSE – Bad Simple English – the culturally destructive language employed for international business communication.

In her analysis of international marketing textbooks, Kelly-Holmes demonstrates how these manuals, which are supposed to argue in favour of international communication and cultural exchange, often imply that "multilingualism is perceived as a chaotic, dangerous and bewildering prospect" while English is often described as the norm, the rescuer that "brings harmony to the world" (2010, p. 196). They also imply that English is the best language of choice when communicating abroad. Moreover, Kelly-Holmes shows a general tendency to judge language professionals negatively and to expect non-native speakers to be unable to understand jargon or complex sentences. Her observations are relevant to our purposes because the positivity towards monolingualism contributes to the image of English as a destructive, *frankensteinian* language.

One more element that needs to be brought into the picture is that of the spread of corporate discourse to other social domains, illustrated by Mautner who theorizes that "the more powerful the agents behind a text or text type are, the more likely they are to be emulated through accommodative acts" (2010, p. 223). As corporate discourse is associated to the English language, it is easy to see how the two are related and tend to influence spheres that do not appear to benefit from a corporate-like approach. Mautner analyses the examples of higher education, public administration, religion and the personal sphere, and shows how discourse in these domains has been influenced by a corporate and business approach, thus creating hybrid discourse which at times is at odds with the domain itself.

In a similar fashion we can add environmentalism to the list. The first environmentalist movements in the 1960s and 1970s were characterized by a kind of typical oppositional discourse. Nowadays environmentalism is growing increasingly institutionalized, and the discourse changes accordingly. Environmentally-friendly activities seek legitimization and attempt to achieve it also through discourse strategies influenced by the corporate realm. This attitude generates contradictions that hinder communication, as activities that can be considered genuinely environment-friendly may present themselves as excessively business-oriented and thus might be perceived as environmentally dangerous.

9.3 The myth of English as the language of freedom, rebellion and modernism

As Crystal (2003) points out, English has played a prominent role as the language of the quality press (p. 91), of BBC radio – "inform, educate and entertain" (p. 96) – and the language of protesters all over the world, using English to address a global audience. Moreover, he argues, as the lyrics of Bob Dylan, Bob Marley, John Lennon, Joan Baez and others spread around the world, in many countries English became a symbol of freedom, rebellion and modernism for the younger generation (p. 103).

It is possible to find examples of this use of English on the Italian websites of environmental movements such as Greenpeace. The Anglicisms employed convey a feeling of modernity and remind the reader of young people's jargon. They can thus be considered part of a persuasion strategy and their analysis aims to focus better on the use of Anglicisms in such texts, bearing in mind the evident contrast with the role played by Anglicisms in the business-oriented communications described above.

Moreover, it was in Britain that the *Magna Carta* – the historical document still celebrated as the first example of the recognition of human rights – was drawn up. The *Carta* was accompanied by another much less celebrated but also extremely important document known as *The Charter of the Forest*. As Linebaugh explains:

> Whereas the first charter concerned, for the most part, political and juridical rights, the second charter dealt with economic survival. Historians have always known the Charter of the Forest existed but many of its terms – for example estovers, or subsistence wood products – seem strange and archaic, and have prevented the general public from recognizing its existence and understanding its importance.
>
> (2008, p. 6)

According to Cowell (2012), "affinity for the landscape, especially in its 'traditional' forms, runs deep in British collective culture and psyche" and a "new commons" approach potentially offers solutions to today's environmental problems. Danny Boyle's opening ceremony for the 2012 Olympic games in London foregrounded the ancient love of British people towards the countryside and the cultural clash brought by industrialization: the success that his show obtained proved that this contrast is still extremely relevant.

9.4 Illegitimate greenwashing and legitimate green marketing

The economic system of the so-called Western countries has been the object of considerable discussion in recent years, especially after the collapse of Lehman Brothers in 2008, which is considered the event that triggered the current economic crisis: the capitalist system has often been criticized and considered "unsustainable" in the long run. Given the central role of capitalism and of the globalization of the economy, marketing techniques have often been accused of maintaining and promoting the capitalist *status quo*, and hence have been identified as anti-environmental (el-Ojeili and Hayden, 2006).

Linguists know that the same communication strategy can be employed both for a specific goal and for its opposite. (Conoscenti, 2011, pp. 73–91) As unlimited consumption and the creation of superfluous needs are by definition in conflict with real sustainability – which aims to reduce needs and consumption to the minimum – any business activity seems doomed to some degree of contradiction. (See also Poli, 2011 for a critique of the concept of "sustainable development".)

Following van Dijk's work on persuasion and manipulation (2006), this chapter hypothesizes a continuum between illegitimate greenwash and legitimate green marketing (see also Caimotto and Molino, 2010). The OED definition for 'greenwash' is *misleading publicity or propaganda disseminated by an organization, etc., so as to present an environmentally responsible public image; a public image of environmental responsibility promulgated by or for an organization, etc., regarded as being unfounded or intentionally misleading.* If we were to consider any marketing strategy as unfriendly towards the environment, analysis would hardly be able to enrich the debate. On the contrary, if we agree that some kinds of marketing persuasion can be accepted and sometimes even welcomed by environmentalists, we can then establish with greater clarity which ones can be considered environment-friendly strategies (Grant, 2007).

9.5 Green(wash)ing Anglicization

In the last few years it has been possible to observe, on the one hand, a growth in the public's awareness of and sensitivity towards environmental problems while, on the other hand, a growth in the amount of business activities that try to capitalize on this new sensitivity by marketing themselves as green. Consumers thus need critical tools to distinguish reliable companies from those who are greenwashing. The hypothesis presented here is that the observation of the use of Anglicisms in corporate communication in countries where English is not the L1 – namely countries included in Kachru's (1988) outer circle – can be employed as a critical tool to help the public establish what are the company's green credentials. Such findings can also be employed to observe the discourse strategies enacted and to draw conclusions to be employed also for the observation of green discourse in texts where English is the L1.

The first aspect that needs to be assessed is what is meant by Anglicism and Anglicization. Furiassi, Pulcini and Rodríguez González (2012, pp. 5–10) provide a complete overview of the various possible interpretations and state that "what counts as an Anglicism may be tailored to the scope of the research" (p. 5). Gottlieb's inclusive definition is probably the best suited for our purposes: "any individual or systemic language feature adapted or adopted from English, or inspired or boosted by English models, used in intralingual communication in a language other than English" (as cited in Furiassi et al., 2012, p. 5). According to Kelly-Holmes (2000),

The use of English in intercultural advertising is quite a unique case, since the English language has meaning, use and significance independent of the countries in which it is spoken. Thus, we see its use as a symbol of a national identity, of globalism, of youth, of progress and modernity; at one and the same time, it can bear the properties of pan-Europeanness/Americanness/ globalism.

(p. 76)

Given the wide array of explanations for the use of English in marketing and advertising, Kuppens' (2010, p. 116) categorization provides a good starting point: she explains that reasons can be grouped under three sets: the larger marketing strategy of using the same slogan worldwide in order to have a consistent brand image and cut costs, the creative-linguistic reasons that allow copywriters to create puns, fill lexical gaps and soften taboos, and, thirdly, the reasons related to cultural connotations. On this last point, she agrees with Kelly-Holmes' observation that, in intercultural advertising today, "the use value of languages has come to be obscured by their exchange or symbolic value" (Kelly-Holmes, 2000, p. 71) and works "even (or perhaps especially) because it is not understood" (p. 73).

But Kuppens also adds a new dimension to the analysis, taking into account case studies in which the choice for English cannot be explained by the three sets of reasons reported above, but rather in terms of intertextual reference to specific American and British media genres (p. 129). Among her findings, the most relevant to our purposes concerns the demands these advertisements make on their public: while slogans which are not meant to be understood by their target imply that the consumer will be uncritically attracted by the "luxury value" associated to English, the TV ads analysed by Kuppens draw on their public's ability to link certain features to specific media genres (p. 131). Kuppens also points out that advertisers adopting English do not necessarily celebrate or admire the cultural values associated with it (Kuppens, 2010).

In a similar fashion, Piller (2001) takes into account the case of a non-profit advertisement which is implicitly critical of the use of English in German advertising: the message was created by a non-profit organization lobbying for the interests of bicyclist and the headline was based on a pun: *Rush hour = Rasch aua*. Piller's explanation deserves to be quoted in full:

> The German homophone of rush hour translates literally as 'quick ouch', so that the whole sequence may be read as 'The rush hour quickly leads to injuries'. [...] The ADFC, which promotes bicycle use as an environmentally friendly alternative to automobile use [is] necessarily critical of the wasteful lifestyle of the rich capitalist market economies and consumer societies. [...] The use of the English term intertextually alludes to commercial advertising in which English is used precisely to endorse the values of capitalist market economies and consumer societies. [...] The use of English in the ADFC advertisement manages to point out the harmful consequences of such a lifestyle without specifying them in so many words.

(p. 174)

This oppositional use of English is particularly relevant for our purposes, as it is linked precisely to the kind of capitalistic values that tend to be associated to the English language, but at the same time it exemplifies the apparent contradiction of using the English language in order to criticize the values it is associated with. Coming back to our overall purpose of establishing ways to detect greenwashing through the observation of Anglicization, the studies quoted above appear to confirm the dichotomy between the choice of using English to exploit its "use value", that is, the propositional meaning of the message, and that of exploiting its "exchange or symbolic value" (Kelly-Holmes, 2000, p. 71), thus capitalizing on the consumers' admiration towards English and the fact they are unlikely to understand the message.

Previous research into how English is employed in Italian green marketing has highlighted two relevant characteristics, which require further investigation in order to establish their validity as warning bells against greenwashing: one is the presence of English and Italian together to refer to the same concept, when Italian would suffice, the other one is the use of opaque terminology which does not have a specific semantic equivalent in Italian and is already characterized by opacity in the source language. As explained in Caimotto and Molino (2010), this is often the case with *stakeholder*.

One way of verifying these hypotheses is to observe how English is employed by companies that can actually be considered environmentally-friendly as they offer products or services that are not superfluous and their production methods are not highly polluting. In order to carry out this task, companies who took part in the tenth edition of the Italian fair of sustainable lifestyles were observed. These companies are expected to represent a good selection of business activities whose green credentials are reliable, as the fair is organized by Terre di Mezzo – a non-profit association active in the realm of critical consumption – and the exhibitors had to conform to a long list of requirements which are published on the fair's website (Falacosagiusta, 2013).

9.6 The Italian fair for sustainable lifestyles: a case study

According to previous findings (Caimotto, 2013), a higher prominence of Anglicisms on the companies' websites corresponds to a stronger positive attitude towards capitalism and consumption, which, as explained, is in contrast with a genuinely environment-friendly approach. Thus, the level of prominence of English in the actual fair in Milan and on the websites of the exhibitors can be considered indicative both of the companies' reliability and of the validity of the hypothesis.

During a one-day visit to the fair, on 17 March 2013, a qualitative observation of the stands, leaflets, posters and catalogues was implemented in order to gather the required information and verify the hypothesis presented earlier. No cases of opaque use of English were detected, nor examples of an overwhelming amount or prominence of Anglicisms. The same can be said about the observation of the fair's catalogue. Cases of warning bells against greenwashing were not found: no cases of English and Italian together to express the same concept and no use of significantly opaque terms. On a general level, though, the presence of English was certainly evident and many examples of Anglicization were detected, especially when looking at the titles, logos, and company names. In fact, drawing on the various works on Anglicisms quoted earlier, it was possible to establish four categories to which the Anglicisms detected belong: International, Creative, Oppositional, Professional.

As Kuppens (2010, p. 116) points out, brands often use English in their communication in order to cut costs and be understood beyond their national boundaries. Among the companies that were present at the fair, the Danish furniture company Flexa can be considered an example: their brand name evokes the English "flexibility": a particularly apt association as the company produces furniture items for children that can be modified following the new requirements of a child as s/he grows up. This kind of usage is here labelled "International": the goal of this strategy can either focus on the need to save money or that of influencing the consumer's identity as cosmopolitan.

The Creative category includes Anglicisms that employ English because of the possibilities it opens in terms of bilingual word puns and rhyming: an example is a shop called "SAVE, Scarpe&AccessoriVEg," whose website address is www. saveshop.it. The shop sells shoes and accessories that look like leather ones but are not made out of animal's skin. Reference to the English language is found in the acronym SAVE, the choice of the symbol "&" and that of employing the shortened version of the word "vegan". The adjectives "vegano/a/i/e" exist in Italian, but the English "vegan" is also employed.

We see here a mix of choices in terms of word order: "Scarpe&AccessoriVEg" respects the Italian rule of the adjective following the noun and it is this word order that allows the creation of the acronym. On the contrary, the website address respects the English word order with "save shop." "Shop" though is not the only word they employ, as they also repeatedly refer to the Italian pun "negozio" on their Facebook page ("negozio" is the Italian for "shop" and "ozio" for "otium").

The most interesting example of Oppositional use of English is found on the website of a communication agency called "Smarketing": an opposition which starts from their company name through the addition of the privative affix –s. The number of Anglicisms on their website is very low and on their home page we find an example of Oppositional use:

> Un processo di liberazione: dall'immaginario dell'advertising, dal consumismo coatto, verso la felicità della decrescita, per la comunicazione come bene comune. (www.smarketing.it)
>
> [A liberation process: from the imagery of advertising, from compulsory consumerism, towards the happiness of de-growth, for communication as a common good. (Translations are mine unless otherwise specified)]

Here, "advertising" is clearly employed to attach a negative connotation to the notion by preferring the English to its Italian correspondent "pubblicità," thus evoking a world where the presence of English is excessive. One of the members has also published a book and, in its online presentation, he states explicitly that it is necessary to avoid using words in English just for the sake of it. Another example is found in their courses section:

> Se usi la parola *target*, non c'è verso, cominci a ragionare come un cecchino. (Smarketing.it – italics in the original)
> [If you use the word *target*, no way, you start thinking like a sniper.]

After these observations, it might be surprising to discover that another prominent Anglicism on Smarketing's website is "stakeholder." In the "skills" section the sub-headline states:

> Invece di lavorare *per* il cliente, preferiamo lavorare *con* lui per i suoi clienti e i suoi stakeholder. (Smarketing.it – italics in the original)
> [Instead of working *for* our client, we'd rather work *with* him (sic) for his clients and his stakeholders.]

The explanation for this apparently surprising choice confirms what the present work is trying to demonstrate: the issue of English in green communication is far from simple and it is certainly not possible to identify some specific Anglicisms – like stakeholder – and state they are more likely to signal cases of greenwashing. On the contrary, what is possible is to identify discursive strategies and tools to recognize these. In this case, the syntax of the message does not exploit the advantages of opacity that the term stakeholder offers: here it is not necessary to pin down who the stakeholders actually are, as we do not even know who the agency's customer is. Hence these stakeholders are simply hypothetical and the insertion of the Anglicism can rather be considered a way to share a certain kind of business identity with the potential customers, who are likely to recognize this term as typical of texts concerning their communication strategies.

In fact this last example may rather be included under the label "Professional," i.e. the insertion of an Anglicism as a strategy of identity formation (Piller, 2001, p. 180). In terms of green credentials, the Professional use of Anglicisms is likely to correspond to the most controversial examples of green marketing, as what companies are trying to achieve in this case is the construction of a business-oriented brand image, which – as explained earlier – clashes with the notion of degrowth. In order to observe this kind of strategies and their consequences, another company that was present in the fair is analysed.

A mainly Professional use of Anglicisms was detected on the website "l'Ecolaio." The company comprises two points of sale in Italy, selling various goods made from recycled materials such as stationery items, inkjet cartridges and toys. The aim of their website is to sell their products, to advertise for their shops but also to promote the possibility of opening a new point of sale as a partner. In the section presenting the latter possibility, many examples of Anglicisms can be found: this is likely to be due to their attempt to establish their identity as reliable business partners. Under the section "Apri il tuo store" ("open your store"), consisting of 485 words, the following English words can be found: format (3), Concept Store (2), green, start-up (2), online, network, business (4), partner (2), web, software, store (2).

The fact the company was at the fair is considered here an element of guarantee of its green credentials. In terms of communicative strategies, "format" is an Anglicism that deserves to be analysed. Here are the relevant occurrences within the section addressed at potential partners:

> Non vi proponiamo un format, ma soluzioni flessibili.
> [We do not offer a format, but flexible solutions].
> Scarica il documento completo del Format con tutti gli elementi d'arredo.
> [download the full document of the Format with all the furniture elements]

The reasons why this Anglicism appears significant is that "format" is first employed with a negative connotation, like "advertising" in the case analysed earlier, and later employed to designate the project offered. Once the website visitors open the "Format" document, they may notice that the slogan running under the logo is "qualche cosa in più di un semplice format" [something more than a simple format], still, the title of the Italian brochure is "format stores 2013," where again we notice a mixture of English words and Italian word order. We have here an example of poor framing (Lakoff, 2010), in which different communicative strategies related to Anglicisms have been employed within the same document and for the same word, thus engendering confusion.

This section has demonstrated that companies which can be considered environmentally-friendly tend to avoid an excessive use of Anglicisms and they do not take advantage of the potential opacity an English expression may offer. We must also bear in mind that nowadays the typical target of this kind of company is a "critical consumer" which means that most of these people in Italy are likely to know English quite well, as they tend to be better off and more educated than the average consumer. In fact, this could prove to be a negative element as it might transform itself in a counterproductive strategy of exclusion of those potential consumers that do not belong to the elite niche.

This analysis has also shown that the influence of corporate discourse can be detected in this business sector too, especially when financial matters are being discussed. This use, that we have labelled Professional, is the most controversial one and it is likely to hinder the effectiveness of a company's green credentials, as a consequence of its closer connection to a capitalist approach, as argued in the section that follows.

9.7 Citizens, consumers, stakeholders

In her abstract, Pillar (2001) states:

> A shift from political identities based on citizenship to economic ones based on participation in a global consumer market can be observed, together

with a concomitant shift from monolingual practices to multilingual and English-dominant ones.

(p. 153)

This concept touches upon what links environment-related marketing strategies to the issue of English as a global language. Looking at the way English is employed in environment-related communication in Italian, we have identified two main tendencies, confirmed by previous studies about the use of Anglicisms in advertising. The trend which has been witnessed for a longer time and has been the object of a greater number of investigation is the one that sees English as a language devoid of its use value and employed only for its symbolic value. Kelly-Holmes (2000, p. 70) links this to Marx's concept of the fetishization of commodities. The connection is very apt, as this use of English is representative of a capitalist approach and is found mainly in the marketing of businesses that favour a capitalistic logic.

The other trend concerning Anglicisms goes in the opposite direction: the English employed is meant to be understood by the public, not only for its symbolic value, but also for its communicative potential. On this side of the continuum we also find an oppositional use of Anglicisms in which the English language is employed to evoke and criticize the ideologies associated to the other trend, namely the use of English as an obfuscating strategy.

We can thus identify the first trend as mainly targeted at consumers – not required to understand but rather to accept the companies' marketing a-critically, thus favouring consumption – and the second trend as mainly targeted at citizens – responsible, critical, active interlocutors. This brings us to the connection between these two approaches and the concepts of Commons and Stakeholders.

The notion of the stakeholder has notably been the object of study and debate in the realms of Economics, Politics and Environmentalism. The word itself engenders confusion and, in some cases, has been employed as an obfuscation strategy (Fairclough, 2000). It is a term that belongs to the logic of capitalism and private property, even if its novelty consists in taking into account advantages that can hardly be valued in monetary terms. On the contrary, the concept of common goods refers to citizens and is not meant to be limited to a specific group of people.

In terms of environmental framing, the concept of common goods is much more effective. If we think of a company polluting the territory where its industrial plants are, the local inhabitants will be considered part of the group of stakeholders. If the company follows a stakeholder logic, it will offer the inhabitants something to compensate for the loss of a non-polluted territory (jobs, investments in local infrastructures such as new roads etc.). The inhabitants will probably lack the necessary level of knowledge to contrast the power advantage of the company critically and will find themselves trapped in the no-win choice between health and economic survival.

A logic based on the concept of Commons, on the contrary, will require the company to avoid pollution anyway, as it removes the approach to the territory based on the concept of private property. As the environment is a complex and interconnected system, it is not possible to pollute a limited part of it, respecting the boundaries of private property, and that is why the framing on which environmental communication is based needs to be grounded on a commons logic. As Lakoff (2010) explains, a change in framing requires a long time and a complex process of reframing, but it appears to be the only long-term solution available.

9.8 It's the common good, stupid!

This section aims to demonstrate the theory presented in the previous one by analysing another case, which was the object of debate in Italy. The case is that of Termoindustriale, a company based in Northern Italy and operating in the business of cogenerators, also known as CHP (Combined Heat and Power). The company was attacked together with one of its clients, Citterio, for the construction of a cogenerator in Felino, near Parma, in an area known for its long-established producers of hams and salami. In simple words, a cogenerator generates heat and power and can be fuelled by burning the production rejects on site, which in this case consist in animal oil. While the system is not zero-emission, the fact the rejects are employed to some end rather than wasted and the fact they do not need to be transported elsewhere by oil-consuming motor vehicles results in a lower impact from the environmental point of view.

But the local population and the other producers started a protest against Citterio, in the attempt to impede the construction of the cogenerator. The documents of the local associations lobbying against the construction focused on two aspects: on the one end they presented unreferenced theories about the dangers of the system without comparing them to the current state of things, on the other hand they foregrounded the fact that Citterio would receive EU incentives for building the plant.

Italy is a country where the familiarity of the population with science-related issues remains too low (Corbellini, 2011) and the level of freedom of information is very problematic (RSF, 2013) thus the population is ill-at-ease with the introduction of technological innovations, as lay people lack both the knowledge they would need in order to judge them as well as the possibility of trusting independent and reliable sources of information. As explained by van Dijk (2006, p. 361), this negative consequence of manipulative discourse typically occurs when the recipients are unable to understand the real intentions or to see the full consequences of the beliefs or actions advocated by the manipulator. This means that a company which wants to introduce technical innovations that impact on the environment will need to frame its communication in terms of public benefit for the common good and downplay the economic advantages.

In the case analysed here, the response of the company was to send technical experts to the public debates to explain the technical advantages of the cogenerator, thus falling into what Lakoff labels "the trap of the Enlightment Reason" (2010, p. 72). The detractors, on the contrary, followed an approach which appealed mainly to people's emotions, misnaming the cogenerator by calling it "incinerator" and stating that "it burns the corpses of animals." In terms of communication, Termoindustriale's website clearly followed a business-oriented approach and was addressed at potential customers. As a consequence, the economic advantages of cutting energy costs and receiving EU incentives were foregrounded: they are likely to have inspired part of the criticisms.

The observation of the use of Anglicisms on their website confirmed the previous hypotheses: the English words employed belonged to the business realm (e.g. partner, business, know-how, business plan, mission, problem solver, trading) and to the relevant technical jargon (e.g. Concentrated Solar Power, High Temperature, Medium Temperature, Organic Rankine Cycle).

The website did not try to present the advantages of their technologies from a more emotional point of view. Cases of blatant obfuscation of facts were not detected, but their pages were clearly addressed at potential customers and it was also very clear that they expected their public to be interested in the economic advantages first. This communicative approach is likely to have triggered negative reactions in environmentally-minded citizens. Given the difficulties the company had to face due to the negative reactions of local citizens – eventually declaring bankruptcy in 2014 – this case demonstrates that attention to framing and improved communication strategies may prove more apt than economic advantages when it comes to building an ecological, sustainable economy.

9.9 Conclusions

The overall purpose of this work is not to promote more or less use of English in environment-related discourse, but rather to provide greater knowledge and awareness of its implications. When observing the presence of Anglicisms in Italian texts that deal with sustainability-related topics, two main trends can be recognized: on the one hand Anglicisms appear to be playing an obfuscating role, distancing the target public from the message while, on the other hand, the presence of words in English can be considered an involving strategy, as those who recognize the language or understand its meaning feel included in the elite.

While the obfuscating strategy is likely to signal cases of greenwashing, that is, texts that promote as environmentally-friendly products and services which are not, the involving strategy is likely to be found in those texts where marketing strategies are employed in order to render the communication more effective. It would probably be impossible to draw a neat line to divide potentially manipulative cases of greenwashing from genuine cases of reliable environment-friendly communication. And it is certainly impossible to establish a list of "dangerous words" signalling greenwashing. As a consequence, what appears as the most reliable strategy to distinguish greenwashing-oriented Anglicisms from inclusive ones is to observe how they are employed in different communications and establish the degree of friendliness towards the environment through non-linguistic methods.

Thanks to this approach, a taxonomy of four strategies was proposed International, Creative, Oppositional, Professional – in the attempt to pin down the different goals communicators are pursuing when they insert Anglicisms in their texts. A qualitative approached was deemed to be the most effective, in order to obtain a fine-grained observation of the discursive implications generated by the presence of Anglicisms in Italian environment-related texts. The Professional strategy was found to be the most controversial, the one that companies should be more wary of using. As argued by Lakoff (2010), the whole issue requires complete reframing. Still, greater awareness of the implications of using English as a Lingua Franca in environment-related communication can only help to improve the way in which messages are created and transmitted. As this will influence the general perception of ecological issues and the consequent policies, the result will not be limited to efficient communication but will prove vital for our survival.

Acknowledgements

I would like to thank Professor Martin Solly for his helpful comments on an earlier draft of this chapter as well as the anonymous reviewers.

References

Baker, P., Gabrielatos, C., Khosravinik, M., Krzyżanowski, M., McEnery, T. and Wodak, R. (2008) A useful methodological synergy? Combining critical discourse analysis and corpus linguistics to examine discourses of refugees and asylum seekers in the UK press. *Discourse and Society*, 19(3): 273–306.

Block, D., Gray, J., and Holborow, M. (2012) *Neoliberalism and Applied Linguistics*. London and New York: Routledge.

Boggio, C. (2017) "Pensi che un *bond* sia un agente segreto?" English as a lingua-not-so-franca in Italian financial communication. In Boggio, C. and Molino, A. (eds) *English in Italy. Linguistic, Educational and Professional Challenges*. Milano: Franco Angeli.

Caimotto, M. C (2013) The unsustainable Anglicization of sustainability discourse in Italian green companies. *Textus*, XXVI: 115–26.

Caimotto, M. C. and Molino, A. (2010) Anglicisms in Italian as alerts to greenwashing: a case study. *Critical Approaches to Discourse Analysis across Disciplines*, 5(1): 1–16.

Conoscenti, M. (2011) *The Reframer: An Analysis of Barack Obama's Political Discourses*. Roma: Bulzoni editore.

Corbellini, G. (2011) Italiani, analfabeti della modernità in *Il Sole 24 Ore*. Retrieved from www.ilsole24ore.com/art/cultura/2011-04-16/italiani-analfabeti-modernita-165609.shtml?uuid=Aajpo0GE di Gilberto Corbellini – Il Sole 24 Ore – leggi su http://24o.it/HkAwS.

Cowell, B. (2012) Forests, the Magna Carta, and the "New Commons": Some thoughts for the Forest Panel. *History and Policy*, Retrieved from https://magnacarta800th.com/papers/forests-the-magna-carta-and-the-new-commons-some-thoughts-for-the-forest-panel/.

Crystal, D. (2003) *English as a Global Language*, second edition. Cambridge: Cambridge University Press.

Dieter, H. (2004) Does "Denglish" dedifferentiate our perceptions of nature? The view of a nature lover and language "fighter". In Andreas, G. and Bernd, H. (eds) *Globalization and the Future of German*. Berlin: Mouton de Gruyter, pp. 139–56.

el-Ojeili, C., and Hayden, P. (2006) *Critical Theories of Globalization*. Basingstoke: Palgrave MacMillan.

Fairclough, N. (2000) *New Labour, New Language?* London and New York: Routledge.

Fairclough, N. (2006) *Language and Globalization*. London and New York: Routledge.

Falacosagiusta (2013) Iscrizioni 2013. Retrieved from http://falacosagiusta.terre.it/sezione04/46.

Furiassi, C., Pulcini, V., and Rodríguez González, F. (eds) (2012) *The Anglicization of European Lexis*. Amsterdam: John Benjamins.

Furiassi, C. (2008) What dictionaries leave out: New non-adapted Anglicisms in Italian. In Martelli, A., and Pulcini, V. (eds) *Investigating English with Corpora: Studies in Honour of Maria Teresa Prat*. Milano: Polimetrica.

Furiassi, C. (2017) Pragmatic borrowing: Phraseological Anglicisms in Italian. In Boggio, C., and Molino, A. (eds) *English in Italy. Linguistic, Educational and Professional Challenges*. Milano: Franco Angeli.

Fusari, S. (2012) Anglicisms in the discourse of Alitalia's bailout. In Furiassi, C., Pulcini, V., and Rodríguez González, F. (eds) (2012), *The Anglicization of European Lexis*. Amsterdam: John Benjamins, pp. 325–42.

Grant, J. (2007) *The Green Marketing Manifesto*. England: Chichester; NJ: Hoboken: John Wiley.

Kachru, B. (1988) The sacred cows of English. *English Today* 4: 3–8.

Kelly-Holmes, H. (2000) Bier, parfum, kaas: Language fetish in European advertising. *European Journal of Cultural Studies* 3(1): 67–82.

Kelly-Holmes, H. (2010) Raising language awareness or reinforcing monolingual norms? A study of international marketing textbooks. In Kelly-Holmes H., and Mautner G. (eds) *Language and the Market*. Basingstoke: Palgrave Macmillan. pp. 185–200.

Kuppens, A. (2010) English in advertising: Generic intertextuality in a globalizing media environment. *Applied Linguistics* 31(1): 115–35.

Lakoff, G. (2010) Why it matters how we frame the environment. *Environmental Communication* 4(1): 70–81.

Linebaugh, P. (2008) *The Magna Carta Manifesto Liberties and Commons for All*. Berkley: University of California Press.

Mautner, G. (2010) The spread of corporate discourse to other social domains. In Kelly-Holmes H., and Mautner G. (eds) *Language and the Market*. Basingstoke: Palgrave Macmillan. 215–25.

Phillipson, R. H. L. (1992) *Linguistic Imperialism*. Oxford: Oxford University Press.

Phillipson, R. H. L. (2008) Lingua franca or lingua frankensteinia? English in European integration and globalisation. *World Englishes* 27(2): 250–67.

Piller, I. (2001) Identity constructions in multilingual advertising. *Language and Society* 30: 153 – 86.

Poli, C. (2011) *Mobility and Environment*. Dordrecht: Springer.

RSF (2013) 2013 World Press Freedom Index. Retrieved from https://rsf.org/en/world-press-freedom-index-2013

Stubbs, M. (2001) *Words and Phrases*. Oxford: Blackwell.

van Dijk, T. A. (2006) Discourse and manipulation. *Discourse and Society* 17(3): 359–83.

10 Nature-based tourism in Greek and English with reference to translation

Sofia Malamatidou

10.1 Introduction

Recent years have seen massive changes in environmental practices and attitudes, which have given rise to environmentalism and a major preoccupation with anything that is 'green' or 'eco'. Tourism is no exception to this pattern. Mass tourism, which can be defined as large numbers of people travelling to the same place at the same time and is related to events "brought about by the commoditization of culture and the associated homogenization and standardization of tourist experiences" (Wang 1999, p.352) is typically held responsible for various types of environmental damage (Farsari 2012), such as air and water pollution, damage to flora and fauna, waste, and high energy consumption. To redress the balance between tourism and the environment a number of alternative nature-oriented forms of tourism have been created focusing on nature (such as eco-tourism, rural tourism, and adventure tourism), collectively named nature-based tourism (henceforth NBT).

From the point of view of the travel and tourism industry, this focus on nature, which is defined as "undeveloped resources" that "influence tourism activities as attraction features, settings or pristine areas" (Farrell 2000, p.409), can be understood as activities which occur primarily in nature, and more generally as opportunities to visit natural areas, which have not been exploited by humans. Naturally, the focus will differ based on the specific type of NBT. For instance, ecotourism focuses more on sustainability and education around nature and ecology, whereas adventure tourism is more about activities that take place in nature, like hiking or kayaking. Overall, NBT, as most types of tourism, is a particularly complex phenomenon, and one that has attracted significant attention from academics and tourism practitioners alike.

While NBT has been examined from a number of different perspectives which shed some light into its complexity, the focus of existing research is rarely on language, with very few exceptions (Mühlhäusler & Peace 2001; Stamou & Paraskevopoulos 2004; Trčková 2016). Although it is acknowledged that different geographic locations will focus on different aspects of NBT, which has been shown to be the case for ecotourism (Weaver 2008), the way this is achieved linguistically has so far been neglected. There seems to be a generally accepted, often Anglocentric, view that a homogenous discourse of NBT exists, not taking into account that different destinations, as well as different cultures, are likely to promote different aspects of NBT manifested through different linguistic preferences. Therefore, there is very little understanding of how the discourse of NBT is constructed, and importantly how it differs from the discourse of mass tourism. What is more, no study has ever addressed the issue from a cross-linguistic point of view (the focus is typically on English), or examined the important role that translation might play, given that the vast majority of tourism texts is translated in at least one language. This focus on language is crucial, as it is the means through which destinations reach out to and attract a specific audience by managing (and often inflating) expectations.

The present study aims to address this gap by focusing on the language used on websites promoting NBT (mostly ecotourism and agrotourism) and examining whether a homogenous discourse of NBT exists, which is distinct from mass tourism, or whether differences are observed based on the destination promoted and/or language used, as well as the role that translation plays in negotiating such differences. Two languages are examined, namely English and Greek, given the popularity of the United Kingdom and Greece as tourism destinations. Recognising the important role that nature plays in NBT, the focus is on linguistic items (nouns and adjectives) used to refer to nature. Nature is examined through a number of categories referring to natural resources (e.g. landscape, flora, fauna), allowing for comparisons not only regarding the overall frequency of references to nature but also regarding the distribution of individual categories, where further differences are likely to be observed. In terms of research questions, this study aims to investigate, firstly, the differences between mass and nature-based tourism texts in Greek and English (focusing on each language separately), secondly, the difference between languages in the discourse of mass tourism and that of NBT (focusing on each category of tourism separately), and, finally, how any differences observed are negotiated in translation.

10.2 Nature-based tourism: defining a nebulous concept

Because NBT is a broad and encompassing term, a general lack of agreement exists regarding its definition. Given the complexity of NBT, it is important to examine some of the definitions proposed, and explain how this concept will be understood in this study. It is worth mentioning that originally the term 'ecotourism' was synonymous to NBT (Wallace & Pierce 1996; Weaver 2005; Boo 1990; Lindberg 1991; Aylward & Freedman 1992), but as different types of NBT emerged, ecotourism became more restricted (see later in this chapter), while NBT started being used as an umbrella term.

One of the early definitions is that provided by Valentine (1992, p.108), who defines NBT as "primarily concerned with the direct enjoyment of some relatively undisturbed phenomenon of nature." The defining characteristic here is for tourism to take place in an environment that has not been manipulated by humans, which raises the question of whether a natural park, with its trails and ponds, might be considered as a setting where NBT can take place. Similar definitions focusing on undisturbed or uncontaminated natural areas are provided by Boo (1990) and Ceballos-Lascurain (1987). A more recent definition is offered by Naidoo and Adamowizc (2005, p.160) who argue that NBT is "a non-consumptive activity that should rely on intact natural resources to generate resources." The focus is again on undisturbed nature, but also on consumptiveness as a differentiating criterion between nature-based and mass tourism. Similarly, Fennell (2014, p.40) understands NBT as an overarching term encompassing all those forms of tourism "which use natural resources in a wild or undeveloped form."

When trying to understand the reality of NBT it is useful to also examine the motivation for it, that is, why people engage with this type of tourism and visit destinations that promote themselves as nature-based. Note that I refer to promotion on purpose here, as the reality of a destination compared to how it promotes itself might be different. A large number of motivating factors has been suggested. These include the desire to get back in touch with nature as part of the environmental movement (Lee 1997; Dowling & Fennell 2003; Dekhili & Achabou 2015), as more and more people become aware of the threats posed to the environment, either through coverage in the media, or the importance given to these issues in political and international agendas. Saarinen (2005) interprets the shift from mass to nature-based tourism as a move away from Fordist to post-Fordist production, with young people especially looking for alternatives to mass tourism which can offer non-materialistic values (Teigland 2001). Another reason for engaging in NBT, and in particular adventure tourism, is the range of activities it offers, such as bird-watching, hiking and kayaking (Blamey 2001). Finally, Mehmetoglu (2007) identifies the desire to escape from everyday life as the strongest motivating factor for NBT. An escape to nature is therefore not necessarily guided by an eco-conscience, but simply by a desire to be in a natural setting away from the cityscape of everyday life. Similarly, in their study, Stamou and Paraskevopoulos (2003) identified the dominance of recreational over environmentalist motives of ecotourists. Therefore, in the context of escaping from the city, mass tourism can also offer opportunities for nature-based experiences, for example by going to the beach, which can be considered a nature-based activity (Fennell 2014; Weaver 2008).

The generally broad definitions of NBT, as well as the fact that the motivation for it is often no more than a desire to be away from the city, raises the important question of what exactly qualifies as NBT. This is further complicated by the fact that visitors might behave in an environmentalist way in non-NBT locations and vice versa. The examination of relevant literature is rather inconclusive, with Blamey (2001) arguing that decisions will necessarily be subjective. Therefore, I approach NBT in its widest possible sense and question classifications that exclude elements of NBT from mass tourism. Visiting a destination to enjoy nature and be surrounded by it is an adequate reason to classify an activity or experience as nature-based. For example, it is very common for package holidays – the most typical example of mass tourism – to involve a guided tour to a natural resource, such as a rainforest, and even focus on its ecological value. Alternatively, a visit to a park, even if the only aim is to admire nature and get away from the city, can be considered an example of a nature-based activity, but it is also possible for someone to visit a park for bird-watching, an instance of adventure tourism. The example of the park is a good one because it also questions the idea of undisturbed nature often appearing in definitions of NBT since it is the result of, at least to some extent, human intervention to nature. Therefore, the difference between mass and nature-based tourism does not lie in the presence or absence of nature-based experiences, but in the way and the degree to which these are promoted, not least linguistically.

10.2.1 NBT sub-types: ecotourism and agrotourism

In this study, I will analyse NBT websites which are self-labelled as promoting either ecotourism or agrotourism, and it is therefore important to also examine these two sub-types of NBT in more detail. Out of all available sub-types of NBT, ecotourism has probably exhibited the largest growth, with words like 'eco' and 'green', as synonymous to sustainable, becoming buzzwords in the travel and tourism industry. Despite its significance, ecotourism is notoriously difficult to define, and there are as many definitions of the phenomenon as there are studies. Apart from a natural setting, most studies focus either on sustainable development (e.g. Arnegger et al. 2010; Björk 2000; Fennell 2014), or educational potential (e.g. Blamey 2001; Buckley 1994) as defining features of ecotourism. Some studies combine all three elements in their definition, such as the one provided by Weaver, who defines ecotourism as "a form of nature-based tourism that strives to be ecologically, and economically sustainable while providing opportunities for appreciating and learning about the natural environment or specific elements thereof" (2001, p. 105).

And it appears that other forms of NBT suffer from the same ontological problem, with over 15 definitions proposed for agrotourism and related forms of tourism (e.g. agritourism and farm-based tourism). A comprehensive definition is that proposed by Ollenburg (2006, p. 52), who describes agrotourism as "activities and services offered to commercial clients in a working farm environment for participation, observation or education." Increasingly, however, this type of tourism is used to describe activities which simply take place on farms (Roberts & Hall 2001), such as wine- or cheese-making.

While ecotourism might be expected to belong to NBT, the same might not be said about agrotourism, which seems to take place in areas that can be considered as disturbed by humans (see also Hughes & Chirgwin 1997; Whelan 1991). However, some overlap between agrotourism and ecotourism is recognised in the literature (Robinson 2012) and an examination of various agrotourism websites suggests that agrotourism shares many of the defining features of ecotourism, such as nature, education, and sustainability. For example, Eumelia, an organic agrotourism farm and guesthouse in Greece, states on its website, which is part of the data examined for this study, that they approach nature "with respect, looking at problems from a holistic point of view, keeping in mind that nature is not a single tree but a complex organism" and that they wish to "provide an educational aspect to our products bringing our projects and the people that come into contact with them closer" (Vision page). Both of these statements echo strongly the aims of ecotourism. Importantly, agriculture, in this case, is identified as a physical rather than a financial activity (Philip et al. 2010). Finally, the farm is often the setting where activities take place, or rather the starting point for exploring the rest of the natural landscape surrounding it. Although agrotourism and other types of NBT, like ecotourism, might not focus on the same activities, what they have in common is the preoccupation with nature and natural resources. And it is for this reason that for the purposes of this study I will use the term NBT to also refer to agrotourism.

10.2.2 A homogenous category?

NBT is not an artificially distinct phenomenon (Preece et al. 1995), but rather a type of human activity woven into the fabric of tourism. This adds to the complexity of NBT. To further support this idea, many scholars distinguish different levels of engagement with nature. For example, Blamey (2001, p.7) makes the distinction between popular and classical ecotourism, which can be extended to NBT more generally. According to him, classical ecotourism involves "the small-scale, personalised and hence alternative nature of many (…) experiences." Popular ecotourism, on the other hand, can be large-scale and thus share some distinguishing features with mass tourism. Weaver (2005) also makes a similar distinction between hard (active, deep) and soft ecotourism (passive, shallow). He argues that softer-path ecotourists are likely to restrict their activities to a small percentage of area in parks, whereas hard-path ecotourists are more likely to explore deeper. Similarly, Acott et al. (1998) distinguish between deep and shallow ecotourism. They distinguish between an extreme form of shallow ecotourism and mass tourism in the way the experience is promoted, that is in terms of references to nature and nature-based activities. This is also further supported by Kontogeorgopoulos (2004) who argues that ecotourism and mass tourism are linked in a symbolic level, with ecotourism often functioning as part of the mass tourism industry, which has helped the former survive financially. Finally, according to Weaver (2001), it is possible to talk about mass ecotourism, and he recognises that ecotourism can be seen as a variant of mass tourism. The same can be true about NBT, which can spill over into other forms of travel, like mass tourism.

What this discussion aims to show is that tourism categories are often arbitrary and fuzzy, with NBT being a particularly nebulous concept, and much more heterogeneous than what might originally appear (Arnegger et al. 2010). Importantly, we should not treat nature-based tourists as a homogenous ideology-oriented single unit. Differences are likely to be observed both within and across cultural groups. In other words, the concept of NBT should not be considered universal, and we should be open to the possibility that it will be interpreted and manifested differently across languages. This is why it is important to examine how NBT is manifested linguistically across languages and through translation, and how it differs from mass tourism, especially given the fact that NBT is becoming increasingly important, and it is claimed to be the fastest growing sub-sector of the tourism industry (Buckley & Sommer 2000; Gray et al. 2003; Coria & Calfucura 2012; Dowling & Fennell 2003; Fennell 2014; Hawkins & Lamoureux 2001). Therefore, in this study, I will focus on the preoccupation with nature and natural resources as a defining characteristic of NBT, and examine how it is manifested linguistically, focusing on one its main features, that is natural resources.

10.3 Natural resources

Definitions of NBT provided earlier revolve around the notion of natural resources as one of the defining characteristics of this type of tourism. One of the earliest and most well-known accounts of resources is that provided by Zimmerman (1951). His understanding of resources can be summarised by the phrase "resources are not; they become" (Zimmerman 1951, p. 15). They are subjective, relative and static, and only acquire importance based on human needs. Human activity is, therefore, what turns various elements into resources (e.g. oil). The idea that the classification of natural resources might depend on human interpretation is also present in studies of NBT. For example, Cassells and Valentine (1990) identify different ways in which tourism can engage with nature: experiences that depend on nature, those that are enhanced by nature, and finally those for which a natural setting in incidental. Valentine (1992) offers the example of swimming, arguing that if a visitor's main interest is to go for a swim, the natural setting is not important, and therefore incidental. However, whether or not the natural setting is incidental depends on a number of factors, and it requires knowledge of the visitor's motivation. For instance, in Valentine's example, if a visit to the swimming pool does not constitute an attractive alternative, then the natural setting cannot be considered entirely incidental. Such models are difficult to implement when large amounts of data are examined. What is needed is a classification which relies less on subjectivity, and which can help guide the linguistic analysis of tourism texts.

Such a classification is offered by Chubb and Chubb (1981), who distinguish between developed and undeveloped resources in relation to the tourism sector. Developed resources are those that facilitate the use of an area, such as motorways, and buildings. Conversely, undeveloped resources are those available in their raw form and can be found in both urban and rural settings. This classification serves as a better analytical framework for examining natural resources, also because it acknowledges the focus on wild or undeveloped nature, which is also frequently found in definitions of NBT. Chubb and Chubb (1981) identify seven types of undeveloped categories which are relevant to the tourism sector:

- Geographic location
- Climate and weather

- Topography and landforms
- Surface materials
- Water
- Vegetation
- Fauna

While this is a very good starting point for the categorisation of nature-related references in tourism texts, this classification also presents certain interpretive limitations. A case in point is the category of water. According to Chubb and Chubb (1981), a lake belongs to the water category, but it could also belong to that of topography and landforms. Equally, the categories of geographic location, and topography and landforms present a significant amount of overlap and thus can be combined. Therefore, I propose a slightly different categorisation, which is presented in detail in Table 10.1.

It is important to include two additional categories that do not exist in the original model: human-made and generic. The human-made category aims to capture a relatively recent trend in NBT, which is often neglected in the literature, that is agrotourism. A generic category is also needed for all those items that cannot be easily assigned to any of the other categories, with *nature* being the most typical example.

10.4 Corpora and methods

The present study uses methods from the discipline of corpus linguistics, which relies on the analysis of large electronic collections of texts (i.e. corpora) to find and reveal patterns in language. Specifically, it relies on corpus triangulation (Malamatidou 2018), a corpus approach for the examination of translation-related phenomena, which shares some similarities with a mixed-methods approach where different types of data and/or different methodologies are combined. Corpus triangulation is defined as "the combination, in an integrated manner, of multiple (two or more) corpus values and/or attributes from one or more corpus variables and/or the

Table 10.1 Categorisation of natural resources

Landscape	The visible structure of the surface of the earth (e.g. mountain, lake, sea)
Climate and weather	Phenomena and processes related to the atmosphere (e.g. rain, sun, wind)
Natural material	Materials derived from nature (e.g. rock, sand, wool)
Flora	Plant life (e.g. trees, flowers, bushes)
Fauna	Animals, both wild and domesticated (e.g. horse, beaver, eagle)
Human-made	Human interaction with and dependence from nature, the use of nature to directly support human activities (e.g. farm, garden, agriculture)
Generic	Anything else related to nature and the environment (e.g. nature, environment, earth)

use of (two or more) corpus analysis techniques in one study of a single phenomenon" (Malamatidou 2018, p. 34). Based on this definition, there are two types of corpus triangulation: data and methods. On the one hand, corpus data triangulation refers to the combination of multiple corpora, based on one or more corpus aspects of corpus design (e.g. type of texts, languages, etc.). This is informed by the Values-Variables-Attributes (VVA) typology of corpora (Malamatidou 2018). On the other hand, corpus method triangulation is defined as the combination of two or more corpus analysis techniques and can be further divided into within-method, where different quantitative methods are combined, and between-method, where quantitative and qualitative methods are combined. It is also possible to use both data and method triangulation, resulting in multiple triangulation. In this study, I will use both corpus data and between-method triangulation. By using this triangulation approach, a clearer picture of the phenomenon can be obtained, which might not have been possible otherwise, especially for complex phenomena that need to be studied from a variety of angles, as is the case in this study.

In what follows, I will present the corpus design and methodology, and explain how triangulation is achieved in each case. Before I do this, it is important to remind the reader of the research questions this study aims to address, which will guide the triangulation process. The present study aims to investigate, firstly, the differences between mass and nature-based tourism texts in Greek and English (focusing on each language separately), secondly, the difference between languages in the discourse of mass tourism and that of NBT (focusing on each category of tourism separately), and, finally, how any differences observed are negotiated in translation.

10.4.1 Corpus design

Based on the research questions, the corpus examined in this study needs to consist of the following elements: (a) websites promoting mass tourism and NBT, (b) websites in Greek and English, and (c) websites including translated and non-translated texts (from Greek into English). Based on these elements, a corpus of some 400,000 words is created for the purposes of this study, consisting of five components (Table 10.2).

Corpus data were collected in the period from June 2017 to August 2018. For the mass tourism texts, data are collected from official tourism websites, as they are considered to be the most representative of this type of discourse. For the NBT texts, data are collected from websites promoting specific businesses, as sufficient data could not be collected from websites promoting NBT more generally (similar to the mass tourism ones) in either Greek or English. Where possible, an attempt was made to include entire websites. Based on the corpus components identified above, corpus data triangulation is achieved by combining texts in different subcorpus configurations (Table 10.3). This specific corpus design allows for detailed comparisons across a number of corpus components.

There is naturally some overlap across subcorpora, which might share the same corpus components. This might appear as repetitive in terms of analysis.

Table 10.2 Breakdown of the corpus of mass and nature-based tourism texts

Component	Website	No of words
English non-translated mass tourism texts	Visit England	39,185
	Visit Wales	58,886
	Visit Scotland	39,221
	Subtotal	**137,294**
Greek non-translated mass tourism texts	Visit Greece	39,712
	Discover Greece	59,863
	Incredible Crete	48,594
	Subtotal	**148,196**
English non-translated NBT texts	Wild Days Holidays	7,550
	Bamff Estate	4,864
	Wheatland Farm	8,633
	Gorge View Cottage	4,174
	Subtotal	**25,221**
Greek non-translated NBT texts	Agroville: The Ecotourism Experience	5,499
	Enagron Ecotourism Village	5,982
	Eumelia Organic Agrotourism Farm & Guesthouse	5,740
	Agroktima Traditional Guesthouse	2,956
	Subtotal	**20,177**
English translated NBT texts	Agroville: The Ecotourism Experience	5,313
	Enagron Ecotourism Village	5,837
	Eumelia Organic Agrotourism Farm & Guesthouse	5,434
	Agroktima Traditional Guesthouse	3,083
	Subtotal	**19,667**
	Total	**387,370**

However, this assumed overlap offers a comprehensive picture of the phenomenon studied, that is, whether a homogenous discourse of NBT exists, which is distinct from mass tourism, or whether differences are observed based on the destination promoted and/or language used, as well as the role that translation plays in negotiating such differences.

Corpus triangulation is achieved in two main ways (for ease of reference, Table 10.4 summarises the VVA typology of corpora).

Firstly, by combining attributes from the texts variable, i.e. different genres: mass and nature-based tourism. Secondly, values from the corpus type variable are combined: comparable and parallel. Specifically, two types of comparable corpora (one of non-translated texts in different languages, and one of translated and non-translated texts in the same language) and one type of parallel corpus are combined. There is also a combination of languages values, that is, monolingual and bilingual comparable corpora, and attributes, that is, Greek and English non-translated texts. Similarly, there is also a combination from the text variable (translated and non-translated texts). Yet, these combinations are tied to corpus type and, as a result, are secondary in achieving triangulation.

Table 10.3 Corpus compilation

Subcorpus	Components	No of words
A. a comparable, monolingual (Greek), synchronic (2017–18) corpus of non-translated texts (from the categories of mass tourism and NBT)	non-translated Greek mass tourism texts non-translated Greek NBT texts	168,346
B. a comparable, monolingual (English), synchronic (2017–18) corpus of non-translated texts (from the categories of mass tourism and NBT)	non-translated English mass tourism texts non-translated English NBT texts	162,515
C. a comparable, bilingual (Greek-English), synchronic (2017–18) corpus of non-translated texts (from the category of mass tourism)	non-translated English mass tourism texts non-translated Greek mass tourism texts	330,861
D. a comparable, bilingual (Greek-English), synchronic (2017–18) corpus of non-translated texts (from the category of NBT)	non-translated English NBT texts non-translated Greek NBT texts	45,398
E. a comparable, monolingual (English), synchronic (2017–18) corpus of translated and non-translated texts (from the category of NBT)	non-translated English NBT texts translated English NBT texts	44,888
F. a parallel, bilingual (Greek-English), synchronic (2017–18) corpus (from the category of NBT)	non-translated Greek NBT texts translated English NBT texts	39,844

Table 10.4 VVA (Variables, Values, and Attributes) typology of corpora (Malamatidou 2018, p.46)

	Corpus Variables			
	Type	Languages	Time	Texts
Corpus Values	Parallel Comparable Reference	Monolingual Bilingual Multilingual	Synchronic Diachronic	Translated Non-translated
Corpus Attributes	-	Specific languages	Specific time spans	Specific genres

Each research question is answered with the examination of two subcorpora, totalling three stages of analysis. Specifically, subcorpora A and B are examined to establish the differences between mass and nature-based tourism first in Greek and then in English (Research Question 1), with each subcorpus focusing on a different language. Subcorpora C and D are examined to establish any differences between languages in relation to mass tourism and NBT discourse between languages (Research Question 2). This is different from the first stage of analysis where the focus is on text types rather than languages. Finally, subcorpora E and F are examined to inspect the role of translated texts and their relationship to respective texts in the target language, as well as to their source texts (Research Question 3). Only if differences are observed in the first and second stage of analysis, the last two subcorpora are examined, which is evidence of the sequential approach to corpus data triangulation. It is worth stressing here that this approach, where the answer to one research question points to the next, is evidence that triangulation occurs in an integrated manner.

10.4.2 Corpus analysis

Corpus analysis mostly involves the identification of all linguistic items that refer to natural resources in the subcorpora examined. The focus is on nouns and adjectives, and not on adverbs or potentially verbs, as, based on pilot studies, nouns and adjectives have been found to express concepts related to natural resources more clearly and with a higher frequency. For the analysis of data, Sketch Engine, a corpus-processing toolkit (Kilgarriff et al. 2014), was used. The identification of nouns and adjectives was facilitated by the fact that the subcorpora have been part-of-speech (POS) tagged, using the modified English TreeTagger for English data, and the INTERA POS tagset for Greek data, which are the default POS taggers offered by Sketch Engine for each language. For each linguistic category, a frequency list was generated. The lists were manually examined to remove any instances of nouns or adjectives not referring to natural resources. For instance, this process included the examination of concordance lines for polysemous words, like *branch* or *bank*.

Corpus method triangulation is achieved by combining two types of descriptive statistics, namely raw and normalised frequencies, and three types of inferential statistics, namely statistical significance, Bayes Factor (BIC), and effect size. These constitute the combination of different quantitative methods in the analysis of subcorpora, therefore achieving within-method triangulation. Specifically, descriptive statistics are recorded in detailed tables for each corpus component to allow for comparisons across subcorpora. Normalised frequencies are reported, as these allow for comparisons across corpus components, and are calculated based on the ration of occurrence per 1,000 words, which is typical in corpus-based studies. Statistical significance is calculated by employing Rayson's (n.d.) statistical significance (log-likelihood) calculator, which also calculates Bayes Factor. Effect size is measured using percentage difference (%DIFF), which indicates the proportion of difference between normalised frequencies. The different types of statistics employed allow for more valid conclusions to be reached, making sure that where differences are observed they are meaningful and important.

To measure statistical significance and Bayes Factor, a null (H_0) and an alternate hypothesis (H_1) are also required. The H_0 is that any differences observed are due to chance, while the H_1 is that the differences found can be attributed to a factor other than chance – for example, different linguistic conventions employed in different languages or text types. Therefore, statistical significance is used to calculate whether or not any observed differences in the frequencies across corpus components and subcorpora are due to chance, that is, whether or not the H_0 can be refuted, while Bayes Factor is used to calculate the probability that this might be the case. For example, an observed difference might be found to be statistically significant, i.e. not likely to be due to chance, but this likelihood might be high or low, which is what Bayes Factor measures. Pragmatically, a statistically significant difference is one where the probability (p) is smaller than 0.05, while very strong evidence against the H_0 is obtained when Bayes Factor is larger than 10.

10.5 Results

10.5.1 Mass versus nature-based tourism: a cross-linguistic analysis

The first research question, that is whether and to what extent there are any differences between mass tourism and nature-based tourism across languages, is addressed by examining comparable monolingual subcorpora of non-translated texts from the categories of mass and nature-based tourism in both Greek and English. The answer to this question will reveal whether the two languages exploit references to nature in a similar manner to distinguish between the two types of tourism. In other words, whether there are any meaningful differences between the two types of texts in terms of how frequently they refer, and by extension, promote nature. Tables 10.5 and 10.6 summarise the results for Greek and English respectively (normalised frequencies are reported per thousand words). For the inferential statistics, where important differences are observed, cells are highlighted.

Starting with Greek, overall 7,942 (53.60 per 1,000 words) nouns and adjectives referring to nature have been identified in mass tourism texts and 1,439 (71.32 per 1,000 words) in NBT texts. Inferential statistics indicate that this difference is statistically significant ($p < .05$) and that there is very strong evidence against the H_0 (BIC = 80.94). In terms of percentage difference, the proportion is moderate, that is, 28.37%. The fact that NBT texts rely more on nature for the promotion of a destination is to be expected since existing literature on NBT frequently stresses the key role that nature plays in defining this type of tourism. At first sight, and as far as the total frequency of references to nature is concerned, there is a clear distinction between Greek mass and nature-based tourism texts, with nature holding a more prominent place in the latter.

Examining the different categories of natural resources in more detail, it is evident that a number of categories have contributed to this pattern, with the exception of the categories of climate and weather, and natural material. For the latter two categories, we cannot refute the hypothesis that any differences observed are due to chance based on inferential statistics. The categories of human-made and generic natural resources stand out, as the percentage difference between mass and nature-based tourism texts is over 100. A special mention is also required for the category of landscape. Here the reverse pattern is observed, with NBT texts

Table 10.5 Mass vs nature-based tourism in Greek

Category	Mass tourism		NBT		Log-likelihood	BIC	%DIFF
	Raw freq.	Normalised freq.	Raw freq.	Normalised freq.			
Landscape	4,435	29.93	422	20.91	p<.05	42.65	35.48
Climate & Weather	231	1.56	40	1.98	p>.05	-10.17	23.73
Natural Material	666	4.49	84	4.16	p>.05	-11.59	7.63
Flora	1,066	7.19	268	13.28	p<.05	58.87	59.50
Fauna	513	3.46	165	8.18	p<.05	66.56	81.10
Human-made	170	1.15	105	5.20	p<.05	111.16	35.48
Generic	861	5.81	355	17.59	p<.05	245.42	100.68
Total	7,942	53.60	1,439	71.32	p<.05	80.94	28.37

Table 10.6 Mass vs nature-based tourism in English

Category	Mass tourism		NBT		Log-likelihood	BIC	%DIFF
	Raw freq.	Normalised freq.	Raw freq.	Normalised freq.			
Landscape	2,968	21.62	103	4.08	p<.05	471	136.50
Climate & Weather	148	1.08	18	0.71	p>.05	-8.96	41.34
Natural Material	498	3.63	97	3.85	p>.05	-11.72	5.88
Flora	466	3.39	143	5.67	p<.05	14.18	50.33
Fauna	1,219	8.88	355	14.08	p<.05	41.46	45.30
Human-made	655	4.77	171	6.78	p<.05	3.61	34.81
Generic	563	4.10	187	7.41	p<.05	32.29	57.52
Total	6,517	47.47	1,074	42.58	p<.05	0.84	10.86

employing 30% fewer nouns and adjectives referring to nature compared to mass tourism texts.

Focusing on English tourism texts, overall there are 6,517 (47.47 per 1,000 words) nouns and adjectives referring to nature in mass tourism texts, and 1,074 (42.58 per 1,000 words) in NBT texts. Unlike Greek texts, English NBT texts refer, and by extension promote, nature less frequently than mass tourism texts. This difference, although not very high (10.86%), is found to be statistically significant, although the evidence against the H_0 is not strong at all (BIC = 0.84). Based on the quantitative evidence, it can be concluded that English mass and nature-based tourism texts refer to nature with approximately the same frequency, and therefore a clear distinction between the two types of texts cannot be easily made.

In terms of individual categories, a very interesting pattern seems to emerge regarding English texts. Leaving the categories of climate and weather, and natural material aside, which, similar to Greek tourism texts, do not demonstrate any statistically significant differences, the reason why English NBT texts refer less frequently to nature compared to mass tourism ones can be solely attributed to the landscape category. The percentage difference, in this case, is 136.50, which is one of the highest proportions identified throughout the entire corpus examination. This corresponds to mass tourism texts referring to nature 429% more frequently than NBT texts. To further illustrate this with an example: the second most frequent noun in this category is *island*, but while this noun appears with a frequency of 1.21 per 1,000 words in mass tourism texts, its frequency is only 0.20 per 1,000 words in NBT texts, which corresponds to a percentage difference of 142.86.

The reason why both Greek and English mass tourism texts refer more frequently to landscape compared to NBT texts is unclear, but it might have to do with the fact that NBT websites are mostly focused around a single destination and its surrounding area, while mass tourism websites tend to promote larger areas, in some cases the entire country. Therefore, it is more likely for mass tourism websites to mention a wider range of landscape elements, such as beaches, islands, and mountains. This is evidence of the fact that the destination might often dictate promotion.

The first stage of analysis reveals that while there are differences between mass and nature-based texts in Greek, with the latter referring to nature more frequently, a similar pattern is not observed in English, where the two text categories seem to behave similarly. The category of landscape exhibits some interesting patterns, which suggests that this particular category will need to be studied more closely, not least when it comes to translation. Based on these results, it might be concluded that Greek tourism professionals understand the difference between mass and nature-based tourism better than their English-speaking colleagues. While it is not possible to exclude this possibility, this is a rather naïve reading of the results. A more valid explanation might be that nature plays a more central role in the promotion of Greece as a tourism destination, and, therefore, this role is exploited more explicitly in tourism texts. To confirm this assumption, it is necessary to conduct a cross-linguistic study of mass and nature-based tourism texts, which is the next stage of analysis.

10.5.2 Mass tourism and NBT: Greek versus English

During this stage of analysis, the same corpus components are examined as those in the previous stage, but combined differently to form new subcorpora. Specifically, the aim is to examine whether and to what extent there are differences between languages in the discourse of mass tourism and that of NBT, and therefore each subcorpus focuses on a text type (mass tourism vs NBT) rather than on a language, as was the case in the previous stage of analysis. Specifically, two comparable bilingual (Greek-English) subcorpora of non-translated texts are examined, one focusing on mass tourism, and the other on NBT.

Results will help understand in more depth how the discourse of mass and nature-based tourism differs across languages and will complement the results obtained from the previous stage of analysis. This is also a necessary stage before examining translation, as it will highlight any differences between the two languages. Table 10.7 summarises the results for the mass tourism category, while Table 10.8 those from the NBT category.

For the category of mass tourism, important differences are observed between the two languages. Overall, Greek texts refer to nature more frequently than English texts, (53.60 and 47.47 per 1,000 words respectively). Results from the log-likelihood test reveal that this difference is statistically significant (p < .05), while Bayes Factor confirms that there is very strong evidence against the H_0 (BIC = 40.48). However, the overall difference is fairly small, that is, 12.13%. Results from the examination of the comparable subcorpus confirm the assumption that Greek texts have the tendency to refer to and promote nature more frequently, as Greece is popular for its beaches and islands.

Some interesting and varied results are obtained when examining the individual categories of natural resources. Specifically, as also observed in the previous stage of analysis, the categories of climate and weather, and natural material do not reveal any meaningful differences between English and Greek. For these two categories, although results from the log-likelihood test indicate that the differences are statistically significant (p < .05), there is not sufficient evidence against the H_0 based on Bayes Factor (BIC < 10). Given their overall low frequency of occurrence, these categories do not have a strong influence on the total proportions. From the remaining five categories, three (i.e. landscape, flora, and generic) are more frequent in Greek than English, and two (fauna and human-made) follow the reverse pattern. In some cases, the difference is quite high, namely for the categories of flora, fauna, and human-made. In both languages, the category of landscape is the most frequent, which is further evidence as to its significance. The promotion of nature in the two languages relies on highlighting different aspects of nature, and while the focus is mostly on landscape, Greek texts also employ nouns and adjectives referring to flora, as well as those belonging to the generic category, while English texts show a preference for the categories of fauna and human-made.

A varied pattern is also observed when examining NBT texts. Overall, once again Greek texts refer to nature more frequently than English (71.32 and 42.58

Table 10.7 Greek and English mass tourism texts

Category	Greek		English		Log-likelihood	BIC	%DIFF
	Raw freq.	Normalised freq.	Raw freq.	Normalised freq.			
Landscape	4,435	29.93	2,968	21.62	p<.05	179.00	32.24
Climate & Weather	231	1.56	148	1.08	p<.05	-0.01	36.36
Natural Material	666	4.49	498	3.63	p<.05	0.65	21.18
Flora	1,066	7.19	466	3.39	p<.05	185.32	71.83
Fauna	513	3.46	1,219	8.88	p<.05	340.10	87.84
Human-made	170	1.15	655	4.77	p<.05	329.96	122.30
Generic	861	5.81	563	4.10	p<.05	29.62	34.51
Total	7,942	53.60	6,517	47.47	p<.05	40.48	12.13

Table 10.8 Greek and English NBT texts

Category	Greek		English		Log-likelihood	BIC	%DIFF
	Raw freq.	Normalised freq.	Raw freq.	Normalised freq.			
Landscape	422	20.91	103	4.08	$p < .05$	274.95	134.96
Climate & Weather	40	1.98	18	0.71	$p < .05$	3.46	94.42
Natural Material	84	4.16	97	3.85	$p > .05$	-10.44	7.74
Flora	268	13.28	143	5.67	$p < .05$	60.90	80.2
Fauna	165	8.18	355	14.08	$p < .05$	24.40	53.01
Human-made	105	5.20	171	6.78	$p < .05$	-6.09	26.38
Generic	355	17.59	187	7.41	$p < .05$	86.44	81.44
Total	1,439	71.32	1,074	42.58	$p < .05$	155.14	50.47

per 1,000, respectively), but in this case the difference between the two languages is significantly higher, that is 50.47%. Considering also the findings from the previous stage of analysis, it is clear that Greek texts rely more heavily compared to English texts on the central role that nature plays in NBT discourse. This might be related to the fact that Greece is associated with natural beauty, as mentioned earlier, and therefore tourism professionals are more sensitive towards this topic.

As with mass tourism texts, the examination of individual categories reveals a certain variety. Once again, the categories of climate and weather, and natural materials do not demonstrate sufficient differences between the two languages. Given that these two categories have so far across all stages of analysis not demonstrated any differences, it is safe to assume that these are not identified as strongly associated with nature by professionals responsible for the production of tourism texts, or, at least, not as elements that are relevant to NBT. A non-statistically significant difference was also observed for the category of human-made natural resources. This can be explained by the fact that aspects of agrotourism, such as farms, play an important role in the conceptualisation of NBT in both languages. For instance, *farm* (and its Greek equivalents αγρόκτημα and φάρμα), the most frequent noun in the human-made category, is used with approximately the same frequency in English and Greek NBT texts (1.70 and 1.88 per thousand words, respectively). Of the remaining four categories, similar patterns are observed as with mass tourism texts, namely the categories of landscape, flora, and generic are more frequent in Greek than English NBT texts, while the reverse pattern is observed for the category of fauna, which is also the most frequent category in English. In Greek texts, the category of natural resources with the most occurrences is landscape. It is also interesting to note the very high difference (134.69%) in the landscape category between Greek and English NBT texts. Once again, the category of landscape seems to play a key role in the pattern observed.

What these findings reveal is that the language used in NBT texts depends primarily on the destination promoted, rather than the text-type, with very few similarities observed between the two languages for the same category of tourism texts. However, we should not exclude the possibility that some differences, especially those related to the preference for different categories of natural resources also depend on language and culture, in line with the argument proposed by Urry and Larsen (2011) that the tourist gaze, that is how we interpret the world around us and which elements of it we seem to prioritise, is informed by our age, social, and cultural group. Therefore, it is not just a case of Greece being promoted differently to the United Kingdom, with more references to nature. Greek and English texts also have different linguistic and cultural preferences, focusing on different aspects of nature. Given these important differences between Greek and English, it is interesting to examine how they are negotiated in translation.

10.5.3 NBT in translation

The final question that this study aims to answer is whether and to what extent English translated NBT texts differ from mass tourism texts, and what is the role of the Greek source texts, ultimately exploring the position of translated NBT texts. For this reason, a comparable, monolingual (English) subcorpus of translated and non-translated NBT texts is examined, as well as a parallel, bilingual (Greek-English) subcorpus of NBT texts. Table 10.9 presents results from the comparable subcorpus, and Table 10.10 the results from the parallel subcorpus.

Starting from the comparable subcorpus, some differences can be observed between English translated and non-translated NBT texts. Overall, translated texts make higher use on nouns and adjectives related to nature than non-translated ones (56.95 and 42.58 per 1,000 words, respectively). This difference, although not very big (28.88%) is found to be statistically significant (p < .05), and there is very strong evidence against the H_0 (BIC = 35.54). Translated NBT texts in English, it would appear, do not meet the communicate conventions of the target language, referring to nature significantly more frequently than non-translated ones. However, results from the previous stages of analysis also revealed that the frequency with which tourism texts refer to nature is most likely destination-specific, with texts promoting Greece relying more on it. It will be possible to confirm this hypothesis, once the parallel subcorpus is examined.

Regarding individual categories, no significant difference is observed for the categories of climate and weather, natural materials, and human-made. This finding confirms the observation made in the previous stage of analysis that, climate and weather, and natural materials are not strongly associated with nature by professionals responsible for the production of these texts, and this is probably why no difference is observed when comparing a number of corpus components. Conversely, references to the human-made category have been found to be equally frequent in Greek and English non-translated NBT texts, which might also explain the similarities between English translated and non-translated texts. Differences are observed for the categories of landscape, flora, fauna and generic natural resources. For all these categories, the difference between English translated and non-translated texts is statistically significant (p < .05), and there is very strong evidence against the H_0 (BIC > 10). However, some categories exhibit a larger difference than others, with the largest difference observed for the landscape (146.12%) and fauna (108.63%) categories. Once again, these two categories seem to stand out. Specifically, although the former is more frequent in translated than non-translated texts the reverse pattern is observed for the latter, with non-translated texts relying on this category more frequently than translated ones. Considering also the results obtained from the previous stages of analysis, landscape seems to play a key role in texts promoting Greece, while fauna in texts promoting the United Kingdom. This can also be associated with the attention that different cultures pay to natural elements and the idea of the tourist gaze (Urry & Larsen 2011).

The analysis of the parallel subcorpus reveals that, overall, Greek source texts refer to nature more frequently than English target texts (71.32 and 56.95 per 1,000 words respectively). This difference, which is however not very big (22.41%) is statistically significant (p < .05) and there is very strong evidence against the

Table 10.9 Translated and non-translated English NBT texts

Category	Translated		Non-Translated		Log-likelihood	BIC	%DIF-F
	Raw freq.	Normalised freq.	Raw freq.	Normalised freq.			
Landscape	301	15.30	103	4.08	p<.05	146.12	115.79
Climate & Weather	21	1.07	18	0.71	p>.05	-9.13	40.45
Natural Material	86	4.37	97	3.85	p>.05	-9.96	12.65
Flora	259	13.17	143	5.67	p<.05	58.30	79.62
Fauna	84	4.27	355	14.08	p<.05	108.63	106.92
Human-made	105	5.34	171	6.78	p>.05	-6.93	23.76
Generic	264	13.42	187	7.41	p<.05	28.61	57.71
Total	1,120	56.95	1,074	42.58	p<.05	35.54	28.88

Table 10.10 Source (Greek) and target (English) NBT texts

Category	Source Texts		Target Texts		Log-likelihood	BIC	%DIFF
	Raw freq.	Normalised freq.	Raw freq.	Normalised freq.			
Landscape	422	20.91	301	15.30	p<.05	6.77	30.99
Climate & Weather	40	1.98	21	1.07	p>.05	-5.05	59.67
Natural Material	84	4.16	86	4.37	p>.05	-10.49	4.92
Flora	268	13.28	259	13.17	p>.05	-10.58	0.83
Fauna	165	8.18	84	4.27	p<.05	14.21	62.81
Human-made	105	5.20	105	5.34	p>.05	-10.56	2.66
Generic	355	17.59	264	13.42	p<.05	0.61	26.89
Total	1,439	71.32	1,120	56.95	p<.05	21.53	22.41

H_0 (BIC = 21.53). Target texts are situated between Greek source texts and respective NBT texts in English. This indicates that perhaps an attempt is made to negotiate the differences in NBT discourse between Greek and English, especially those aspects which have been found to be language-specific.

The examination of the individual categories reveals that the category of fauna is primarily responsible for any differences in the overall proportion of references to natural resources between source and target texts, as it is the only category where the difference is both statistically significant ($p < .05$) and there is very strong evidence against the H_0 (BIC = 14.21). Linguistic items belonging to the category of fauna have been found to be preferred by English NBT texts, and one might expect that more frequent references to animals might be found in the English target texts following this pattern. However, the opposite pattern is observed. For the remaining six categories, even if there is a statistically significant difference, the evidence against the H_0 is not very strong, and therefore, there is not sufficient evidence to refute the possibility that any difference observed has occurred by chance. However, it is worth noting that the category of landscape could also be responsible to some extent for translated texts standing between Greek source texts and respective English target texts, although there is not sufficiently strong evidence against the H_0 (BIC < 10), which means that we cannot refute with certainty the hypothesis that this difference might be due to chance.

Therefore, at first sight, it might appear as if translated texts make an attempt to negotiate differences, but this attempt does not appear to be informed by existing patterns in the target language, but rather by a translation approach which seems to contradict such patterns. Overall, target texts are very close to their source texts, with the exception of the fauna category, and potentially to some extent that of landscape, although results are inconclusive in the case of the latter. It is possible that translators recognise that tourism promotion is mainly destination-specific, not realising that linguistic and cultural preferences might also be in play. This raises the question of whether translators realise how exactly the promotional function of tourism texts is achieved, and importantly whether they recognise the promotional potential of nature and how it might be achieved across languages, especially in the case of NBT.

10.6 Discussion

The aim of this study has been to examine, focusing on references to natural resources, whether a homogenous discourse of NBT exists, which is distinct from mass tourism, or whether differences are observed based on the destination promoted and/or language used, as well as the role that translation plays in negotiating such differences. It has been found that nature, as a concept which can have promotional value in tourism discourse, is deployed differently in English and Greek tourism texts. Greek relies much more on references to nature to distinguish and promote NBT as different from mass tourism, a pattern that has not been observed in English, where the linguistic items referring to natural resources are used with a similar frequency between mass and nature-based tourism. This can be associated with the fact that, compared to English, Greek tourism texts, irrespective of which category they belong to, refer to nature much more frequently, suggesting that nature is perceived as having a stronger promotional value for Greece. This might be explained by the general perception of Greece as a country with ample natural beauty and as a destination for beach and island holidays (compared to, for instance, a city break). This does not mean that English texts ignore the importance that nature plays in NBT, but rather that they might have other ways of highlighting its central role, for instance, by expressing active engagement with it, for which, however, a different type of analysis is required. Therefore, although according to literature on NBT, we might expect nature to be referred to more frequently in texts promoting NBT irrespective of the language in which they are written and the destination they promote, this study reveals that this dependency on nature is primarily destination-specific, but also to some extent language-specific, as evidenced by the distribution of the categories of natural resources.

Translated texts seem, at least at first sight, to negotiate such differences to some extent by striking a reasonable balance between their Greek source texts and respective NBT texts in English. In a way, translators seem to recognise that the destination, i.e. Greece, needs to be promoted more heavily through nature, but also that English genre conventions suggest that this reliance on nature should be mitigated to some extent. The negotiation of differences in translation is, however, not observed for all categories of natural resources, rather it seems to be driven solely by the category of fauna and landscape, although inferential statistics are inconclusive in the case of the latter. The category of fauna, although found to be preferred in English than Greek texts in both mass and nature-based non-translated tourism texts, is used less frequently in translated texts even compared to the Greek source texts. In this case, there is no clear evidence of a justified translation approach, which might explain the very low frequency of linguistic items associated with fauna in English target texts. For the category of landscape, linguistic items belonging to this category have been found to be more frequently used in mass tourism texts compared to NBT ones, and more frequently in Greek than in English tourism texts. Translated NBT texts stand in between relying on these linguistic items more frequently than respective English texts and less frequently than Greek source texts. A similar pattern has not been observed for any of the other categories.

Therefore, while translators might recognise that the focus on nature relies on the destination being promoted, they seem to have missed the fact that some linguistic differences are also present regarding the distribution of individual categories. It is possible, although quantitative data are inconclusive, that translators identify landscape as the most representative category of nature, and therefore it is the only category where a justifiable translation approach is observed to an extent. This would mean that translators do not fully recognise that nature can be expressed through many other categories of natural resources, which play a significant role, particularly in NBT. Also, given the prominence of the landscape category in mass tourism texts, it seems that translators of NBT texts prioritise elements more representative of mass tourism, therefore approaching NBT texts in more or less the same way as they might approach mass tourism ones.

These findings have significant implications for both the tourism and travel industry, and translation studies. Comparing different categories of tourism texts across languages reveals that aspects that might intuitively be considered universal, to the extent that they might define an entire category of tourism, can be realised differently according to the destination they are promoting and the language in which they are produced. This is the case even for tourism performed in Europe, which is generally considered as a homogenous category (Weaver 2008). The point raised about the language in which a text is produced is associated with the audience which it addresses and how this audience perceives the outside world (or at least how tourism professionals assume it perceives it). Research conducted by Urry and Larsen (2011) suggests that people belonging to different cultures and social groups see and interpret new places differently, which means that the perception of a destination is subjective (Mayo & Jarvis 1981). Consequently, the results of this study challenge the universality of NBT as a category and its distinguishing features and call for more research into what constitutes NBT and how it differs from mass tourism, taking into account cross-linguistic differences and arguments about the subjectivity of destination interpretation.

Translation studies also needs to reflect on its own practices. Results suggest that translators might not fully recognise the significant role nature might plays for a specific audience in a specific language. We need translation to engage more with reflective practice and create stronger links with some of the specialised fields in which it plays a central role, such as tourism. This will help raise awareness among translation professionals regarding the key characteristics of each text type and how they can be approached. At the same time, the present study highlights the fundamental role that translation plays in tourism promotion, and how easily a different interpretation and presentation of a destination through translation might be achieved. Translation, therefore, needs to claim its role in promotional language, and raise its profile in that regard, not least for the benefit of the tourism industry. This is the responsibility of translation studies, but equally of other disciplines, like tourism and travel, that need to start paying careful attention to how translation practice affects their products and services. Only then we will be able to have a fruitful interdisciplinary dialogue, which will result in better cross-cultural communication.

The final point that needs to be raised about the implications of this study is related to similarities observed between English mass and nature-based tourism texts regarding references to nature. As already stated, it should not be assumed that English texts ignore or do not recognise the important role that nature plays in NBT. It is highly possible that English texts highlight this role through other means, for instance through verbs expressing direct involvement with nature. However, it is also possible that what has been seen as a clear-cut distinction a couple of decades ago, that is mass versus nature-based tourism, is no longer so obvious, with references to nature infiltrating more and more frequently mass tourism texts in recent years, at least in the Anglophone world. In that sense,

NBT might not be considered as a relatively distinct category of tourism anymore (which to an extent echoes Weaver's [2001] argument regarding mass ecotourism), with mass tourism and a soft/passive/shallow NBT significantly converging. This raises the question of how 'alternative' NBT is today. It is necessary that tourism categories identified 20 or more years ago are frequently reviewed to access their relevance for today's tourism and travel industry. At the same time, more research is needed into how exactly nature is conceptualised and presented in different categories of tourism. However, we should not exclude the possibility that certain linguistic choices might be insufficient for promoting the desired aspects and principles of NBT, and therefore offer evidence towards 'greenwashing' (Donohoe & Needham 2006; Weaver 2005; Honey 2008). But, it is only by understanding better how these choices operate across languages that we can reach more valid conclusions about the extent to which their use is appropriate or not.

It should be noted, of course, that this study also has certain limitations. The corpus examined, although specialised, is still rather small, and any generalisations should be made with caution. A larger corpus, with a larger NBT component, needs to be studied, consisting of more languages, which will help reveal how heterogeneous NBT is as a category, and to what extent certain preferences are destination- or language-specific. Additionally, the validity of the results relies exclusively on the accuracy of the POS taggers, which even for the English POS tagger is expected to be around 90% (Horsmann et al. 2015). This is not ideal, but equally, it is impossible to conduct a similar analysis manually. To avoid similar issues of reliability, future studies might consider examining in more depth the most frequent (e.g. 10 or 20) items under each category, which can be captured with the use of a frequency list, and without the need for POS tagging. Additionally, a more in-depth analysis will reveal whether engagement with nature can be expressed not only by the frequency with which a text refers to it but also by other means, such as, for example, the verbs used to accompany nouns associated with natural resources.

10.7 Conclusion

The idea of alternative tourism is not something new, but our understanding of how different alternative tourism really is compared to mass tourism is very limited, while there is also a commonly accepted view that different types of tourism show universal characteristics. This has resulted in a significant amount of effort dedicated to defining concepts like NBT, but not realising that the difficulty associated with defining these might be related to the fact that different cultures might interpret and perform these types of tourism differently. This study has addressed this gap by offering for the first time insight from a cross-linguistic comparison of mass and nature-based tourism texts. As a result, it has questioned the way in which NBT has been perceived until now and also highlighted the important role that translation plays in its conceptualisation. It is necessary that more studies of this type, which focus on linguistic aspects, are conducted on tourism texts, which will allow us to reveal not only how different cultures look at the world, but also help the tourism and travel industry understand better how language might have an impact on destination promotion.

References

Acott, T., Trobe, H. & Howard, S., 1998. An evaluation of deep ecotourism and shallow ecotourism. *Journal of Sustainable Tourism*, 6(3), pp. 238–53.

Arnegger, J., Woltering, M. & Job, H., 2010. Toward a product-based typology for nature-based tourism: A conceptual framework. *Journal of Sustainable Tourism*, 18(7), pp. 915–28.

Aylward, B. & Freedman, S., 1992. Ecotourism. In B. Groombridge, ed. *Global Biodiversity*. London: Chapman & Hall, pp. 413–15.

Björk, P., 2000. Ecotourism from conceptual perspective. *International Journal of Tourism Research*, 2, pp. 189–202.

Blamey, R.K., 2001. Principles of ecotourism. In D. Weaver, ed. *The Encyclopedia of Ecotourism*. Wallingford: CABI Publishing, pp. 5–22.

Boo, E., 1990. *Ecotourism: The Potentials and Pitfalls*. Washington: WWF.

Buckley, R., 1994. A framework for ecotourism. *Annals of Tourism Reserach*, 21(3), pp. 661–65.

Buckley, R. & Sommer, M., 2000. *Tourism and Protected Areas: Partnerships in principle and Practice*. Gold Coast: Queensland.

Cassells, D. & Valentine, P., 1990. Recreation management issues in tropical rainforest. In *Institute of Tropical Rainforest Studies Workshop 1*, Townsville: James Cook University.

Ceballos-Lascurain, H., 1987. The future of "Ecotourism." *Mexico Journal* (January), pp. 13–14.

Chubb, M. & Chubb, H., 1981. *One-Third of Our Time? An Introduction to Recreation Behaviour*. New York: John Wiley.

Coria, J. & Calfucura, E., 2012. Ecotourism and the development of indigenous communities: The good, the bad, and the ugly. *Ecological Economics*, 73, pp. 47–55.

Dekhili, S. & Achabou, M.A., 2015. The perception of ecotourism: Semantic profusion and tourists' expectations. *RIHME*, 19(5), pp. 3–20.

Donohoe, H. & Needham, R., 2006. Ecotourism: The evolving contemporary definition. *Journal of Ecotourism*, 5(3), pp. 192–210.

Dowling, R. & Fennell, D., 2003. The context of ecotourism policy and planning. In D. Fennell & R. Dowling, eds. *Ecotourism Policy and Planning*. Oxford: CABI Publishing, pp. 1–20.

Eumelia Organic Agrotourism Farm & Guesthouse. Available at: www.eumelia.com/farm/vision-philosophy/ [last accessed November 2018].

Farrell, T., 2000. Nature. In J. Jafari, ed. *Encyclopedia of Tourism*. London: Routledge, pp. 409–10.

Farsari, I., 2012. Sustainable tourism policy in North Mediterranean destinations. *Journal of Hospitality Marketing and Management*, 21(7), pp. 710–38.

Fennell, D., 2014. *Ecotourism* 4th edn. London: Routledge.

Gray, P. et al., 2003. The socioeconomic significance of nature-based recreation in Canada. *Environmental Monitoring and Assessment*, 86, pp. 129–47.

Hawkins, D. & Lamoureux, K., 2001. Global growth and magnitude of ecotourism. In D. Weaver, ed. *The Encyclopedia of Ecotourism*. Wallingford: CABI Publishing, pp. 63–72.

Honey, M., 2008. *Ecotourism and Sustainable Development: Who Owns Paradise?* 2nd edn. Washington: Island Press.

Horsmann, T., Erbs, N. & Zesch, T., 2015. Fast or accurate? A comparative evaluation of PoS tagging models. In *Proceedings of the International Conference of the German Society for Computational Linguistics and Language Technology*. Duisburg-Essen: University of Duisburg-Essen, pp. 22–30.

Hughes, K. & Chirgwin, S., 1997. Ecotourism: The participant's perceptions. *Journal of Tourism Studies*, 8(2), pp. 2–7.

Kilgarriff, A. et al., 2014. The sketch engine: Ten years on. *Lexicography*, 1, pp. 7–36.

Kontogeorgopoulos, N., 2004. Conventional tourism and ecotourism in Phuket, Thailand: Conflicting paradigms or symbiotic partners? *Journal of Ecotourism*, 3(2), pp. 87–108.

Lee, C.-K., 1997. Valuation of nature-based tourism resources using dichotomous choice contingent valuation method. *Tourism Management*, 18(8), pp. 587–91.

Lindberg, K., 1991. *Policies for Maximising nature Tourism's Ecological and Economic Benefits*. Washington, DC: World Resources Institute.

Malamatidou, S., 2018. *Corpus Triangulation: Combining Data and Methods in Corpus-Based Translation Studies*. London: Routledge.

Mayo, E.J. & Jarvis, L.P., 1981. *The Psychology of Leisure Travel: Effective Marketing and Selling Travel Services*. Boston: CBI.

Mehmetoglu, M., 2007. Nature-based tourism: A contrast to everyday life. *Journal of Ecotourism*, 6(2), pp. 111–26.

Mühlhäusler, P. & Peace, A., 2001. Discourses of ecotourism: The case of Fraser Island, Queensland. *Language & Communication*, 21, pp. 359–80.

Naidoo, R. & Adamowicz, W., 2005. Economic benefits of biodiversity esceed costs of conservation at an African rainforest reserve. *Proceedings of the National Academy of Sciences of the United States of America*, 102(46), pp. 16712–16.

Ollenburg, C., 2006. *Farm Tourism in Australia: A Family Business and Rural Studies Perspective*. Gold Coast: Christian-Albrechts University of Keil and Griffiths University.

Philip, S., Hunter, C. & Blackstock, K., 2010. A typology for defining agritourism. *Tourism Management*, 31, pp. 754–58.

Preece, N., Van Oosterzee, P. & James, D., 1995. *Two Way Track – Biodiversity Conservation and Ecotourism: An Investigation of Linkages*. Canberra: Department of the Environment, Sports and Territories.

Rayson, P., n.d. *Log-Likelihood Calculator*. Available at: http://ucrel.lancs.ac.uk/llwizard. html [last accessed February 2018].

Roberts, L. & Hall, D., 2001. *Rural Tourism and Recreation: Principles to Practice*. Cambridge: CABI Publishing.

Robinson, P., 2012. Agritourism. In P. Robinson, ed. *Tourism: The Key Concepts*. London: Routledge.

Saarinen, J., 2005. Tourism in the Northern Wilderness: Wilderness discourses and the development of nature-based tourism in Northern Finland. In C.M. Hall & S. Boyd, eds. *Nature-Based Tourism in Peripheral Areas: Development or Disaster?* Clevedon: Channel View Publications, pp. 36–49.

Stamou, A. & Paraskevopoulos, S., 2004. Images of nature by tourism and environmentalist discourses in visitors books: A critical discourse analysis of ecotourism. *Discourse & Society*, 15(1), pp. 105–29.

Stamou, A.G. & Paraskevopoulos, S., 2003. Ecotourism experiences in visitors' books of a Greek reserve: A critical discourse analysis perspective. *Sociologica Ruralis*, 43(1), pp. 34–55.

Teigland, J., 2001. The effects on travel and tourism demand from three mega-trends: Democratization, market ideology and post-materialism as cultural wave. In W. Gartner & D. Lime, eds. *Trends in Outdoor Recreation, Leisure and Tourism*. Wallingford: CABI Publishing, pp. 37–46.

Trčková, D., 2016. Representations of nature in ecotourism advertisements. *Discourse and Interaction*, 9(1), pp. 79–94.

Urry, J. & Larsen, J., 2011. *The Tourist Gaze 3.0*. London: SAGE Publications.

Valentine, P., 1992. Review: Nature-based tourism. In B. Weiler & C. Hall, eds. *Special Interest Tourism*. London: Belhaven Press, pp. 105–27.

Wallace, G. & Pierce, S., 1996. An evaluation of ecotourism in Amazonas, Brazil. *Annals of Tourism Reserach*, 23(4), pp. 843–73.

Wang, N., 1999. Rethinking authenticity in tourism experience. *Annals of Tourism Reserach*, 26(2), pp. 349–70.

Weaver, D., 2001. Ecotourism as mass tourism: Contradiction or reality? *Cornell: Hotel and Restaurant Administration Quarterly*, 42, pp. 104–12.

Weaver, D., 2005. Comprehensive and minimalist dimensions of ecotourism. *Annals of Tourism Reserach*, 32(2), pp. 439–55.

Weaver, D., 2008. *Ecotourism* 2nd edn. Milton: John Wiley.

Whelan, T., 1991. Ecotourism and its role in sustainable development. In T. Whelan, ed. *Nature Tourism: Management for the Environment*. Washington: Island Press, pp. 3–22.

Zimmerman, E., 1951. *World Resources and Industries*. New York: Harper.

Index